Time Travel: A Reality in Mesoscopic Physics

P. Singha Deo · Kanchan Meena

Time Travel: A Reality in Mesoscopic Physics

Springer

P. Singha Deo
Sector III
S. N. Bose Centre
Salt Lake, Kolkata, West Bengal, India

Kanchan Meena
Sector III
S. N. Bose Centre
Salt Lake, Kolkata, West Bengal, India

ISBN 978-981-96-1138-6 ISBN 978-981-96-1136-2 (eBook)
https://doi.org/10.1007/978-981-96-1136-2

This Springer imprint is published by the registered company Springer Nature Singapore Pte Ltd.
The registered company address is: 152 Beach Road, #21-01/04 Gateway East, Singapore 189721, Singapore

P. Singha Deo dedicates this work to his parents.

Kanchan Meena dedicates this work to her parents.

Foreword

Professor P. Singha Deo (PSD) is known to me for a long time, since he has been a Ph.D. student. His supervisor was also well known to me. His student and co-author K. Meena is also known to me and I have personally discussed with her. She is a sincere and methodical student keen for indepth study. Prof. Singha Deo's previous student Dr. Satpathi is also known to me and some of the works reported in this book are in collaboration with her for some of which I have been the referee in peer reviewed journals. I have often been in the review committees of both these students including Dr. Singha Deo. I got interested in their work about eight years ago and followed them out of curiosity. PSD is working on a very important problem for a long time, which one has to admit is controversial as well as counter-intuitive. The Argand diagrams with little sub-loops, leading to negative partial level densities and time travel, are very interesting. There is also a simple and fresh approach to the issue of the state of a tunneling particle under the barrier. Most mathematical steps have been worked out in the book and the reader, therefore, should be able to follow the logic with ease. They have also pointed out several direct and indirect methods to test their results experimentally. The results are anyway around for quite some time and are deeply rooted in basic physics to warrant a textbook in the subject. The results also seem to lead to many other important problems in physics. I wish the authors best of luck.

Prof. P. K. Panigrahi
(Director) IISER Kolkata
Kolkata, India

Preface

This is a substantial modification over the previous book titled "Mesoscopic Route to Time Travel" by PSD. A major expansion and refinement has been possible due to the contribution of a bright Ph.D. student (KM) who has brought out several ways to experimentally confirm the reality of time travel in mesoscopic systems. Time travel is an extremely fascinating subject in physics and does change our world view. But this book will also introduce a student to several very fundamental problems in physics that can be answered in the mesoscopic regime. Notable among them are (a) state of a tunneling particle under the barrier, (b) a local description of the state of a quantum particle, and (c) reconciliation of the classical and quantum notions of time. The book paves the way to develop an alternate description of quantum mechanics and a different route to unification of classical and quantum laws. Essential differences with textbook statistical mechanics have been pointed out. Major breakthrough in basic science always has a lot to offer in terms of applications and future technologies. That too has been discussed and will be of interest to researchers.

At various stages PSD was involved in collaborations with Dr. U. Satpathi, Dr. S. Mukherjee, Dr. S. Bandopadhyay, etc. He is also indebted to his teachers, Prof. A. M. Jayannavar, Prof. F. M. Peeters, Prof. M. Manninen, and Prof. B. K. Talukdar to name a few. He would also like to extend his courtesy to Prof. P. Panigrahi, Prof. S. Dattagupta, Prof. M. Buttiker, Prof. P. A. Sreeram, Prof. B. C. Gupta, Prof. S. Bhattacharjee, Prof. T. P. Pareek, and Prof. P. K. Ghosh as some of them have provided valuable support and others lively discussions. The author's personal life was greatly enriched by his wife, parents, a host of friends (mostly birders, to specially mention Shyam Ghate, Bhaskar Das, Sanjay Thakur, and Sopan Sen), and of course the lively music of Salil Choudhury and Mark Knopfler that helped him cope with the hardships and frustrations of research. The author has personally experienced that while trained scientists think similar and are more predictable, an intelligent comment by a layperson or even a line in a song requires him to think out of the box.

KM would like to express her gratitude to her parents and siblings. Without their trust and support, she would not have reached her milestones. They give her emotional strength and help her in being focused. Next words are, of course, for her supervisor Prof. Prosenjit Singha Deo who guided her well with a positive attitude in the right direction. She could progress only because of his regular discussions and guidance.

Kolkata, India

P. Singha Deo
Kanchan Meena

Contents

About the Authors

Prof. P. Singha Deo has worked at the S. N. Bose National Centre for Basic Sciences in Kolkata, India, since 1999. He has successfully guided several Ph.D. students leading to their Ph.D. in physics and also mentored a few postdocs. His works are mostly analytical and theoretical, but he has collaborated closely with experimentalists and co-authored many scientific papers with experimentalists. He has consistently worked on the problems elaborated in the book for over 25 years and more. He has also taught several courses in physics at the M.Sc. level. Before joining S. N. Bose Centre, he spent about 3 years in Europe as a postdoctoral fellow. He did his Ph.D. in physics in 1996 from the Institute of Physics, Bhubaneswar, India.

Kanchan Meena joined S. N. Bose Centre, Kolkata as a Ph.D. student under the supervision of Prof. P. Singha Deo in 2019. She submitted her thesis in 2024 and a large part of her Ph.D. work has fueled this book. She has played a proactive role in her research endeavor and she is greatly interested in a career in Physics research. She did her masters in Physics from Vardhaman Mahaveer Open University, Rajasthan, India. Her hobbies include watching classic movies, photography and dancing. She is also interested in popularizing physics among friends and young souls. She has suffered the brunt of the pandemic and fought it bravely to tell the tale and now she wants to move forward.

Abbreviations

1D	1 dimension
2D	2 dimension
3D	3 dimension
AB	Aharonov-Bohm
AC	Alternating current
AD	Argand diagram
BC	Bergers circuit
CSBS	Complete set of basis states
CSOF	Complete set of orthonormal functions
DC	Direct current
DOS	Density of states
EL	Euler-Lagrange
Eq.	Equation
Eqs.	Equations
Fig.	Figure
FSR	Friedel sum rule
GTR	General theory of relativity
HC	Hermitian conjugate
KE	Kinetic energy
LDOS	Local density of states
LPDOS	Local partial density of states
LPT	Larmor precession time
PDOS	Partial density of states
PE	Potential energy
Q1D	Quasi-one dimension
Sc.	Schroedinger
STR	Special theory of relativity
TDS Eq.	Time-dependent Schroedinger equation
TDS	Time-dependent Schroedinger
WDT	Wigner delay time
ZFC	Zermelo-Fraenkel-Choice

Chapter 1
Open Systems

Abstract This chapter gives an introduction to some basic concepts. There is a fundamental theoretical problem in modeling and understanding the origin of dissipation and loss. In mesoscopic physics the primary focus is on quantum transport and thermodynamic properties that originate from flow and fluctuation of charge or current that is intricately connected to dissipation and loss. This chapter will provide a novel approach to this old problem and how it is done for mesoscopic systems. Next we will orient the reader to the von-Neumann approach to quantum mechanics, which is believed to be the correct theory of quantum mechanics, in dealing with unbounded operators that model the continuum. That will show the importance of studying the problem using first quantization and a basic introduction will be given as to how to treat boundary conditions leading to continuity of wave-functions and conservation of currents. The boundary condition approach can naturally deal with the situation when there are evanescent modes in the system that most matrix diagonalization approaches overlook. Such evanescent modes will become very important in later chapters. Also, a typical thermodynamic property, i.e., magnetization of a mesoscopic system due to AB effect will be explained in detail using the continuum approach. Special emphasis will be given to explain the topological nature of the problem. An elementary introduction to density of states and its connection to scattering phase shift will be provided.

Sample fabrication techniques in material physics and low temperature cryogenics have evolved to such an extent that with their help one can conduct physics experiments with samples that are as small as 10 to 100 microns. Such samples have novel features that lead to the development of the subject of mesoscopic physics. First is that the equation of motion for electrons in these systems is that of quantum mechanics, namely the Schrodinger equation. This essentially requires that the inelastic mean free path or phase breaking length of the electrons is comparable with the sample size. At low temperatures, typically milli Kelvin, the inelastic mean free path is greatly increased. Secondly, the relevant length scales that determine the physics of these systems, like the elastic mean free path, or field penetration length, or any correlation length, etc., become comparable to the sample dimensions, and as a result the artificially created competing length scales lead to new and novel phenomenon in these

P. S. Deo and K. Meena, *Time Travel: A Reality in Mesoscopic Physics*,
https://doi.org/10.1007/978-981-96-1136-2_1

systems. This has opened up renewed interest in addressing two central questions in physics. First is that how can we make sense of quantum mechanics and how does it cross over to classical behavior as the sample size grows. Secondly, is it possible to control the quantum fluctuations and use the novel properties of these systems to build devices.

One very intriguing feature of quantum systems is the linear superposition of states which is often stated in the form of Schrodinger's cat paradox. It essentially means that a quantum cat can be dead and alive at the same time which is difficult to accept in reality. Another intriguing feature is that quantum mechanics can allow non-local behavior, a good example for which is the Aharonov-Bohm (AB) effect that will be discussed in Sects. 1.2 and 1.3. In this regard, this book will firmly establish that there is a local object in quantum mechanics, namely the local partial density of states (LPDOS), that will show up in present day state of the art mesoscopic experiments, although the axiomatic framework of quantum mechanics denies its existence. These local objects can be also summed up in classical sense to lead to the correct density of states that can be obtained from quantum mechanics and hence the notion of these local states can co-exist with the rest of quantum mechanics. There is a well defined unique time spent by electrons in these local states and not that the electron is partly in a local state and partly somewhere else at the same time. Time spent at local states can be summed up in the classical sense to find the total time spent. This total time spent is also consistent with relativity and can co-exist with it, opening up the possibility of uniting quantum mechanics and relativity at low energies rather than the usual notion that one should try for this unification at high energies.

The key question that the mesoscopic regime addresses is that what exactly is loss or dissipation or decoherence and how it sets into a conserved system. Experiments that probe this aspect are transport properties like resistance, capacitance, inductance, etc. An isolated atom like the hydrogen atom does not allow for a definition of such notions and there is no microscopic theory for such effects. The phenomenological approach of the Landauer conductance formulas has been immensely successful in describing conductance of small systems including that for an atom or molecule and suggests the fact that there could be objects like LPDOS. Such a suggestion of LPDOS coming from a phenomenological theory, however successful may be, will have conceptual issues associated with it which this book will clarify. In some of the ideal samples the Landauer conductance formula has been tested to an accuracy of one part in a billion which is far more accurate than predictions we get from solving Hamiltonian quantum systems. Only other experimental data with similar agreement with theory is the Lamb shift. Another important fact brought out by the Landauer conductance formula is that in mesoscopic systems we cannot define material specific quantities like resistivity, permeability, dielectric constant, etc., as ensemble averaged correlation functions. This is essentially because these material specific parameters are determined by the intrinsic length scales of the material such as correlation length. In mesoscopic systems they face competition from external length scales like sample dimension, curvature of surfaces, etc. But we can define global properties like resistance, capacitance, magnetization, etc., for mesoscopic samples. The philosophy can be extended to study capacitance, inductance, AC (alternating current) response, non-linear response, etc., but for some key unanswered questions and doubts on LPDOS. The issue of loss is related to aging and the arrow of time and that is the reason for these doubts. This book intends to clarify that by showing that time travel is possible or the reversal of the arrow of time is possible and as a result lossy channels can add to increase of coherent current that can be experimentally observed with present day state of the art techniques. Of course the corresponding questions can be asked also for a purely classical system but this book will be dedicated to mesoscopic systems and the theories that have developed that help us understand the origin of loss in quantum systems. In mesoscopic physics, it is much more transparent as to how irreversibility sets in and allows us to treat semi-irreversible systems. In the subsequent section we present a particular way of seeing this coming from the basic postulates of quantum statistical mechanics as outlined in the first 3 pages of Chapter 8 in K. Huang, Statistical Physics, 3rd edition. So there are no additional inputs like the arguments presented by Landauer in his original paper or the arguments given in the book by S. Datta. So we believe that readers who are psychologically set back by the new concepts involved in mesoscopic physics will be able to make themselves more comfortable with our axiomatic approach and proceed. We quote here a paragraph from the book by Dennery and Krzywicki, Dover publications, NY, page 1. "The standard way to circumvent the difficulty of defining fundamental mathematical objects is to formulate a certain number of axioms which are the rules of the game and which form the basis of a deductive theory. The axioms are fashioned upon the intuitive properties of very familiar objects, such as the integers or the real numbers, but once the axioms are adopted, we need no longer appeal to our intuition. In other words, when the rules have been specified, the question of knowing exactly what these objects represent is no longer relevant to the construction of a relevant theory".

Devices like the present day computer or mobile phones or any device that require an electronic circuit use the law of large numbers. This essentially means that current or magnetization or any response of a sample is due to a statistically large number of carriers, say electrons, typically of the order of the Avogadro number. So resistance or capacitance or magnetization is an average response produced by the carriers, the fluctuations are of the order $\frac{1}{\sqrt{N}}$ as we have seen in statistical mechanics text books, and they can be ignored. This creates an upper limit on what we can achieve technologically. We can break this barrier by using mesoscopic principles wherein a few electrons can be used to create currents and responses that will make the devices fast and easy to manage because of low energy dissipation. Thus we will not use averages but then fluctuations that will be quantum in nature will make our devices very unreliable. In this book we will show how evanescent modes can be used to solve this problem for which one has to address the question of what exactly is current and magnetization due to evanescent modes.

The primary development in fabrication technique that has facilitated this subject is described in the book by Datta [1]. It relies on mechanically bringing together two three-dimensional (3D) cubic blocks, one of i-GaAs and the other is n-AlGaAs. Their lattice structure is so close to each other that at the interface a two-dimensional electron gas (2DEG) is formed which is a very high mobility electron gas trapped at the interface due to the nature in which the lattice structure relaxes at the interface. The 3D blocks are then removed and the 2DEG is isolated. One can apply lithography or etching techniques on this 2D film to curve out geometries like an isolated ring or a sample connected to electron reservoirs. The equation of motion for electrons in such systems is just the single particle Schrodinger equation wherein the bare mass of the electrons is replaced by an effective mass. This effective mass incorporates the effect of other interactions that may be present in the system. Through out this book we will focus on such realistic samples and hence when we talk of time travel or local density of states in these systems, then all these effects can be realized in the laboratory. It is to be noted that one can also make such samples with metals and crystalline semiconductors but the physics of those systems will be different in terms of the theory. For example in gold and copper rings the electrons are mostly diffusive and may not be studied using the effective mass single particle Schrodinger equation.

The book by Datta [1] is mainly written for engineering students who intend to numerically apply Landauer two probe, three probe, and four probe conductance formulas wherein one starts from a Hermitian Hamiltonian or a scattering matrix. Such an application oriented book will rely on representing the Hamiltonian in the form of a matrix and apply recursive techniques [2]. So this approach will not lead to fundamental aspects of quantum mechanics in a straight forward approach. One has to look for clever ways of modeling and incorporating the related effects. The correct theory for quantum mechanics was developed by von-Neumann and others and beautifully explained in the YouTube lecture series by Frederic Schuller. Operators in quantum mechanics are not supposed to be Hermitian but are supposed to be self-adjoint. Through out this book we will stick to first quantization or in other words we will solve for the continuous energy or wave vector of a scattered electron that is consistent with the von-Neumann approach. It is known that for large crystalline systems the Hermitian Hamiltonian and diagonalization approaches are valid but at no point we are talking of such large systems with a periodic potential. It is possible that in some cases the continuum model can be mapped to a discrete model with a Hermitian Hamiltonian. For example the effects of evanescent modes in a single channel quantum wire can be mapped to a tight binding stub (to be discussed in Chap. 2) to give some aspects of the physics involved but this inspection is never done in the community. Thermodynamic properties of mesoscopic systems have been particularly ignored in Datta's book and so is the role of evanescent modes, that we believe leads to many new phenomena and will attract a lot of attention in the future. Like S. Datta, we will restrict to electronic properties only because phononic properties in mesoscopic systems are mostly done in analogy with electronic properties and have not yet reached a level of maturity to be presented here in our opinion. Of course the role of phonons on electronic properties has been presented to some extent by S. Datta but that is again an alternate view to what will be also discussed here.

In the first two chapters we have presented two approaches to mesoscopic systems, open and closed. This is a general philosophy in physics wherein one can see the closed system as a special case of an open system or one can see the open system as a special case of the closed system. One encounters this dichotomy formally in several forms like micro-canonical system versus grand canonical system in statistical mechanics, closed intervals versus open intervals in mathematics, conservative systems versus dissipative systems in classical physics, eigen-states versus resonant states in quantum mechanics, etc. In mesoscopic physics this comes in the form of systems coupled to reservoirs versus isolated systems. Essentially in the isolated system we get eigen-energies while for the system coupled to reservoir we get resonant states and one can in fact make the broadening of the resonant states tend to zero by tuning the coupling strength of the system to the reservoir but there are always some standard issues as to whether we can get to the limit of eigen-states. One may just as well ask if in nature we at all get eigen-states. Normally in physics one thinks the closed systems can be exactly understood as everything is known whereas the open systems are those in which a few degrees of freedom are averaged out or ignored and most physical systems are open. In mesoscopic systems this debate has always been very intense because averaging out of some degrees of freedom is against the philosophy of mesoscopic physics. Landauer followed by Buttiker introduced the idea that mesoscopic open systems are coupled to a special kind of electron reservoirs that grasp physical situations remarkably well. Besides at the non-interacting level there are new ingredients in mesoscopic physics like magnetization due to Aharonov-Bohm effect and an even-odd effect called the parity effect. So in the first two chapters we have discussed these in detail. In this book we will adopt the view that open systems are more fundamental than closed systems because of our recent findings that will be highlighted in Chaps. 5, 6, and 7. In the second chapter we will discuss the closed systems corresponding to the open systems in Chapter 1. There we will also discuss a new phase of the electron wave-function that was first reported in a closed system but has dramatic effects in open systems.

If one has to adopt the hardcore approach of discarding definitions that are derived from the Hamiltonian followed by averaging or integrating some degrees of freedom, then the Larmor clock approach resulting in a hierarchy of density of states is suited to treat mesoscopic systems because it can be only defined for a system connected to reservoirs. In this approach one can define local entities like local partial density of states, emissivity, injectivity, etc., that cannot be defined from the Hamiltonian. However, their meanings were not very clear in terms of experiments that can be designed and pose a rather challenging mathematical concept that the Larmor clock approach lead to local descriptions and hidden degrees of freedom in the sense that one can talk of particular electrons that are going from one particular initial state to one particular final state wherein one can even talk of a number at a local point inside the system which is related to the local state of the electron at that point. And there have remained special doubts related to the fact that these numbers may become negative. So in third chapter we have given a detailed account of the formal derivation of the Larmor clock and pointed out the issues associated with it. Our contribution is outlined in Chaps. 4, 5, and 6 where we give strong theoretical support for the Larmor clock. In Chap. 7 we further point out why and how it will manifest in actual experiments. The "why" can be answered from the occurrence of sub-loops in Argand diagrams (AD) of quantum mechanical scattering matrix elements. The Larmor clock formalism may allow us to directly access global properties like capacitance, inductance, AC response, non-linear response, etc. Any other formalism that first requires us to calculate resistivity, dielectric constant, permeability, susceptibility, and other such material specific parameters that depend on intrinsic scales is not what we see as suited for mesoscopic systems for which by definition they do not exist. Of course one can interpret terms in a broader sense but that is beyond the focus of this book. Therefore we will restrict to a regime where not much has been done and there is a huge scope from the technological point of view as well as for basic science.

Before we go to Chaps. 5 and 6 where we solve the pertinent problems associated with the Larmor clock mentioned in the previous paragraph, in Chap. 4 we briefly outline exact mathematical analysis to treat realistic physical systems and in particular cases we present a verification that the Larmor clock is physically and mathematically correct. We also show how the even-odd parity effect can be utilized to reveal this. But of course these are particular cases which are therefore generalized in Chaps. 5 and 6. The analysis in Chap. 4 illustrates the key components of the von-Neumann approach wherein the energy and wave vector can vary continuously while the basis states are infinite but countable. The eigen-energies are obtained from the singularities of the scattering matrix and not from any diagonalization scheme. We will show that mathematically Larmor clock is consistent with the complex plane and the topology of the complex manifold. And physically, the Larmor clock is related to the physical propagation of a non-dispersive wave-packet in time and for such a situation there can be a local time. Negative partial density of states and negative propagation times are possible in real systems around Fano resonances. While the results in Chap. 5 can have a lot of technological applications, Chap. 6 will show the possibility of time travel and that can greatly enhance our basic understanding of the universe we live in. Chapter 7 will discuss several experimental possibilities. In some cases existing experimental data just needs to be reinterpreted and tuned to the problem. In some cases theoretical cross checks and theoretical experiments make the possibilities bright.

Equilibrium thermodynamics: Magnetization and polarization are two of the primary equilibrium thermodynamic properties. The first is the response of a sample to applied magnetic field and the second is response to an applied electric field. One can ask the question if one can define magnetization or polarization of an isolated atom or molecule. The general formalism to study and understand such effects can again be the Larmor clock. It can also become an ideal tool to include the effect of electron-electron interaction that will be briefly discussed in Chap. 5. The relevance of that formalism can be seen in Ref. [4] where we microscopically derive and define quantum capacitance and numerically show that the formalism holds good in presence of electron-electron interactions. In this book we will mostly discuss magnetization for electrons in the effective mass model (as in GaAs-AlGaAs interface) as that leads to some dramatic results to be described in detail in Chaps. 5 and 6. Understanding magnetization in detail will help readers grasp the basic physics that might be extended to other unknown areas in the future. Mesoscopic magnetization is predominantly dictated by interference effects like the Aharonov-Bohm effect and Lorentz's forces play a negligible role. For example mesoscopic disk can exhibit large magnetization due to a few Gauss while a few Tesla of magnetic field is required to yield Hall effect related responses. So we present below a very basic introduction to response of a 1D ring to a magnetic flux penetrating the ring, there being no magnetic field in the ring. This system is a natural choice because it can be treated analytically and the concepts can be clearly explained. For a quantum disk or any other structure the same physics manifests and one can see the Aharonov-Bohm oscillations with a periodicity of Φ_0. But theoretically such systems can only be solved numerically. There is a topological aspect to this problem which is usually not emphasized to beginners and will be explained in Sect. 1.2.

1.1 Transport Current

Consider a typical situation where a DC (direct current) is driven through a small quantum system. In order to approach this problem, consider a 1D scattering problem discussed in textbooks and illustrated in Fig. 1.1. The equation of motion is the 1D Schrodinger equation,

$$-\frac{\hbar^2}{2m_0}\frac{d^2\psi}{dx^2} + U(x)\psi(x) = E\psi(x) \tag{1.1}$$

where m_0 is the electron mass. In regions I and III of Fig. 1.1 where potential $U(x)$=0, solution of this differential equation gives

$$\psi_I = ae^{ikx} + be^{-ikx}$$

$$\psi_{III} = ce^{ikx'} + de^{-ikx'} \tag{1.2}$$

where $x' = x - l$. As mentioned, it is generally believed that the correct axiomatic approach to Quantum Mechanics was developed by von-Neumann and others about a century ago but textbooks in quantum mechanics often do not follow it. As a result the concepts found in textbooks often run into contradictions and a student often takes a lot of time to come to terms with them, unless in the meantime he gets demotivated and quits. Textbooks often assume that the momentum operator is Hermitian and therefore momentum states are well defined. As a result one can solve this kind of scattering problem in 1D by saying that a particle once transmitted never turns back because there is nothing to throw it back or reflect it back. This means that for the 1D scattering problem depicted in Fig. 1.1, in the negative semi-axis one sets a=1, b=r (reflection amplitude), and $c = t$ (transmission amplitude), d=0 on the positive semi-axis. This picture works well when generalized to 3D in certain cases like the physics of particle colliders. Essentially it is taken that scattered states in 3D are of the form $e^{i\vec{k}\cdot\vec{r}}$ in analogy with 1D but ironically, that does not satisfy the 3D Schrodinger equation and the time reversed solutions are also ignored for no reason. These mathematical problems are bypassed in a phenomenological way to lead to a concept of scattering cross section and optical theorem that are practically very relevant. It is not guaranteed that the same phenomenology will describe physical situations in 1D and quasi-1D (Q1D). In fact they do not, there are mathematical inconsistencies and physically not relevant. How such 1D and Q1D physical situations arise will be described in Chap. 4.

Consider the one-dimensional (1D) plane waves in Eq. 1.2, which unlike in 3D are true asymptotic solutions of 1D Schrodinger equation. Observables in quantum mechanics are given by self-adjoint operators and symmetric (symmetric operators are wrongly called Hermitian in confusion with finite dimensional matrices) operators are not enough to qualify. This is specially true for momentum operators and some cases are described by Bonneau et al. [3] including the case of semi-infinite leads. For an elaborate discussion one may see the lectures on quantum mechanics by Frederic Schuller on YouTube. But let us see if the momentum operator $\hat{p}$ is at least symmetric (for the simple minded condensed matter community we will refer to as Hermitian) which is always the initial test for an operator to qualify to describe an observable. To check symmetry, one has to consider an inner product of the form

$$p_{fg} = \langle f | \hat{p} | g \rangle = \int_{-\infty}^{\infty} f^* \hat{p} g dx = \int_{-\infty}^{\infty} f^* \frac{\hbar}{i} \frac{dg}{dx} dx$$

$$= f^* \frac{\hbar}{i} g|_{-\infty}^{\infty} - \int_{-\infty}^{\infty} \frac{df^*}{dx} \frac{\hbar}{i} g dx$$

Usually the first term is set to zero without much inspection and consequences are discussed below. Therefore,

$$p_{fg} = \int_{-\infty}^{\infty} \frac{df^*}{dx} \left[\frac{\hbar}{i}\right]^* g dx$$

Here * means complex conjugation and $g^{**} = g$. Therefore

$$p_{fg} = \int_{-\infty}^{\infty} \left[\frac{\hbar}{i} \frac{df}{dx}\right]^* g dx = p_{gf}^* \tag{1.3}$$

Let us consider the term set to zero, i.e., $f^* \frac{\hbar}{i} g|_{-\infty}^{\infty}$. If f and g are normalizable functions, then they go to zero at $\pm\infty$. But plane waves like the solutions of the Schrodinger equation in the leads (as in Eq. 1.2) are not normalizable. So at least

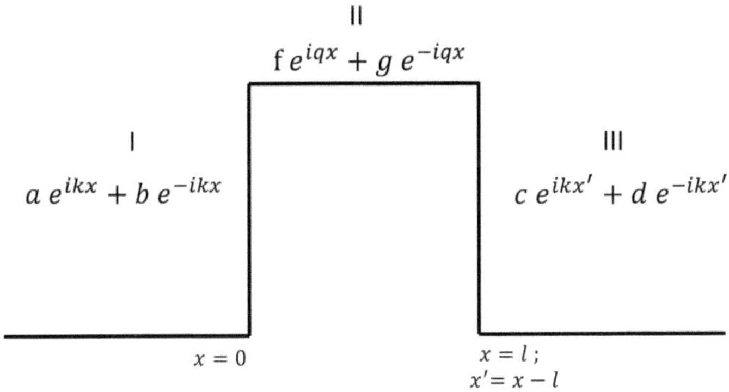

Fig. 1.1 A textbook style scattering problem of a 1D rectangular barrier

for such plane waves in semi-infinite leads we cannot say that $\hat{p}$ is symmetric (or Hermitian). Besides, most normalizable wave-functions are of the form of a standing wave where symmetry does not guarantee a well defined momentum state.

Thus for the situation depicted in Fig. 1.1, we will not use $a = 1$, $b = r$, $c = t$, and $d = 0$ which is how the problem is solved in textbooks. In condensed matter systems we cannot rule out electrons incident simultaneously from the left as well as from the right. As electrons in the leads go far from the scatterer, they react with the environment. Far away from the scatterer a, b, c, and d depend on the coordinates of the external world as well as on time (this time is obviously the classically recorded time at a point where the electron starts interacting with the classical world). Thus according to Eqs. 8.5 and 8.6 in Kerson Huang, we can say

$$|\,\bar{a}\,|^2 = 1, |\,\bar{d}\,|^2 = 1, \overline{ab} = \overline{cd} = \overline{a^*b} = \overline{ab^*} = \overline{cd^*} = \overline{c^*d} = \overline{d^*a} = \overline{a^*d} = 0 \tag{1.4}$$

We will use the conditions in Eq. 1.4 and not $a = 1$, $b = r$, $c = t$, and $d = 0$. This means cross terms like $\lim_{x\to\pm\infty} \bar{f}^* \frac{\hbar}{i} \bar{g} = 0.0$ implying symmetry of momentum operator, expressed by Eq. 1.3. The bars on top like $\bar{a}$ imply averaging over measurement time, and Eq. 1.4 is known as random phase approximation. Obviously we are dealing with a statistical mechanical system in an environment and applying axioms of statistical mechanics. What that means in terms of a mathematically well defined self-adjoint momentum operator is left as an open question for future works at this point. We do not encounter the mathematical problem of showing a self-adjoint momentum operator because we naturally have electrons incident from both directions. In fact k in Eq. 1.2 can be simply interpreted as a wave vector. Landauer's philosophy is essentially the same as that in Eq. 1.4 with a sharp demarcation between the classical world and quantum world which is very realistic and will also be described in detail in Sect. 1.3. But if one is a little alienated by Landauer's arguments then this text book idea can be used. The essential point is that one cannot explain Eq. 1.3 from a purely quantum mechanical viewpoint of Hamiltonian systems. We take the momentum operator to be symmetric for our system shown in Fig.

1.1 but since we have not shown it to be self-adjoint as well, we will proceed without assuming $d = 0$ and instead use Eq. 1.4. We will see that this positions Landauer's phenomenological approach with respect to the axioms.

According to Eq. 1.4, close to the scatterer

$$|\, a\,|^2 = 1 \text{ and } |\, d\,|^2 = 1 \tag{1.5}$$

signifying that they have nothing to do with the scatterer, representing flux of incident particles equally probable from the left as well as the right. The bar on top will often not be explicitly mentioned. Thus we are avoiding setting $d = 0$ by hand. Close to the scatterer b and c will be determined by the scatterer. In Fig. 1.1, the amplitude b in region I can thus be written as

$$b = ar + dt'$$

at $x = 0$. This is because amplitude b at $x = 0$ contains the reflected part of a, i.e., ar and the transmitted part of d, i.e., dt'. Similarly, one can write

$$c = dr' + at$$

at $x' = 0$. Here r' and t' stand for reflection and transmission amplitudes due to incidence from the right. So total wave-function in region III of Fig. 1.1 will be

$$\psi_{III} = dr'e^{ikx'} + ate^{ikx'} + de^{-ikx'}$$

Probability current density J in region III will be (e_0 denotes single electron charge)

$$J = \frac{e_0\hbar}{2im_0}(\psi^*_{III}\nabla\psi_{III} - \psi_{III}\nabla\psi^*_{III})$$

$$= \frac{e_0\hbar}{m_0} Im(\psi^*_{III}\nabla\psi_{III})$$

$$= \frac{e_0\hbar}{m_0} Im\Big[(d^*r'^*e^{-ikx'} + a^*t^*e^{-ikx'} + d^*e^{ikx'})\frac{d}{dx}(dr'e^{ikx'} + ate^{ikx'} + de^{-ikx'})\Big]$$

Using Eqs. 1.4 and 1.5 and simplifying we get

$$= \frac{e_0\hbar}{m_0} Im[(ik)(|\, r'\,|^2 + |\, t\,|^2 - r'^*e^{-2ikx'} + r'e^{2ikx'} - 1)]$$

$$= \frac{e_0\hbar}{m_0} Im[(ik)(|\, r'\,|^2 + |\, t\,|^2 - |\, r'\,|\, e^{-2ikx'-i\eta} + |\, r'\,|\, e^{2ikx'+i\eta} - 1)]$$

$$= \frac{e_0\hbar}{m_0} Im[(ik)(\mid r' \mid^2 + \mid t \mid^2 + (2i) \mid r' \mid Sin(2kx' + \eta) - 1)]$$

where $r' = |r'|e^{i\eta}$. So the current density on the right lead at T = 0 K for $e^{\pm ikx}$ incident wave (i.e., unit incident flux from both sides) is

$$J = \frac{e_0\hbar k}{m_0}[\mid r' \mid^2 + \mid t \mid^2 - 1]$$

or

$$dJ = \frac{e_0\hbar k}{m_0}[\mid r' \mid^2 + \mid t \mid^2 - 1]dn \tag{1.6}$$

The differential current in Eq. 1.6 signify the fact that now instead of unit flux $|d|^2 = 1$ incident from the right (see Eq. 1.5), now we have a flux of $dn \neq 1$ incident from the right. The third term in Eq. 1.6, that is $\frac{e_0\hbar k}{m_0}[-dn]$ is the incident current, wherein unit incident flux from the right, i.e., $|d|^2 = 1$ is being replaced by dn. The sign just signifies the direction and can be dropped. Note that determining current is just a count or measure and so whether $\frac{e_0\hbar k}{m_0}$ can be interpreted as $e_0 v$ where v is a velocity (and $\hbar k$ a momentum) is relevant only if we do experiments that are analogous to photoelectric effect of photons. The quantum mechanical expression for incident current is consistent with a count $e_0\frac{\hbar k}{m_0}dn$ or $e_0 v dn$, probabilistically crossing a point coming from the probabilistic interpretation of wave-function in QM. One may thus write from the third term in Eq. 1.6 that the incident current in an energy interval dE is

$$dJ_{in} = e_0 v dn = e_0 v \frac{dn}{dE} dE = e_0 v \frac{1}{hv} dE$$

where $\frac{dn}{dE} = \frac{1}{hv}$ is the 1D density of states (DOS). And this results in the current through the sample to be (from Eq. 1.6)

$$dJ = \frac{e_0\hbar k}{m_0}(\mid r' \mid^2 + \mid t \mid^2 - 1)\frac{dn}{dE}dE \tag{1.7}$$

Therefore

$$dJ = \frac{e_0\hbar k}{m_0}\frac{\mid r' \mid^2 + \mid t \mid^2 - 1}{hv}dE \tag{1.8}$$

$$= e_0 v \frac{\mid r' \mid^2 + \mid t \mid^2 - 1}{hv}dE = \frac{e_0}{h}[\mid r' \mid^2 + \mid t \mid^2 - 1]dE \tag{1.9}$$

which was also derived by Landauer. Note from comparing Eqs. 1.7 and 1.8 that incident current normalized by $\frac{1}{hv}$ is consistent with the 1D density of states, so that the number of states involved is $\frac{1}{hv}dE$. Essentially $\frac{e_0\hbar k}{m_0}(\frac{-1}{hv})dE$ in Eq. 1.8 can be identified with dJ_{in}, that is the incident current before scattering. While the current after scattering is $\frac{e_0\hbar k}{m_0}\frac{|r'|^2+|t|^2}{hv}dE$. Note that dJ in Eq. 1.9 is therefore zero at zero temperature exhibiting Kirchoff's law and hence the sign convention for the terms. A formal explanation of this sign convention can be seen from Eqs. 1.42 and 1.43 to be presented below. Hence a normalization constant for the wave-function

$$\left(\frac{1}{hv}\right)^{1/2} = \left(\frac{2\pi m_0}{h^2k}\right)^{1/2} \tag{1.10}$$

will automatically take care of the fact that at zero temperature the current at a given energy is carried by the available quantum states which equals the DOS. This basic concept is essential to the understanding of mesoscopic transport and will be used subsequently in various forms. Now at finite temperature [1] a non-equilibrium current can be obtained by taking the temperature on the left different from that on the right modifying Eq. 1.9 to

$$dJ = \frac{e_0}{h}\left[f_r(E) \mid r' \mid^2 + f_l(E) \mid t \mid^2 - 1 f_r(E)\right] dE \tag{1.11}$$

Here the first term is the reflection of electrons incident from the right and so they will come with the temperature of the right reservoir and so on. Here $f_r(E)$ is the Fermi function for the electrons coming from the right reservoir and $f_l(E)$ is that for those coming from the left. The so-called Landauer reservoir will be described in detail when we discuss the system of Fig. 1.3. Alternatively at zero temperature one can take a chemical potential difference between the left and the right. If there is a current through the system then the system can have a resistance which can be quantitatively calculated in the Landauer formalism and that as stated above is one of the greatest successes of theoretical physics. Therefore, Eq. 1.4 leads to the Landauer conductance formula on one hand and the symmetry of the momentum operator on the other.

An incident wave-function in a quantum mechanical scattering set up is rather correctly described by wave-packets and the plane waves in Eq. 1.2 are very correctly the Fourier components making the wave-packet. In the theory of quantum mechanics, the Fourier operator is more basic than the Schrodinger operator, which means the Fourier components of a wave-packet will always satisfy the Schrodinger equation. But not all solutions of the Schrodinger equation are admissible. A solution of the Schrodinger equation that cannot be a Fourier component making the wave-packet is a wrong solution. These solutions are often ignored in textbooks as "nonphysical". In our opinion use of terms like physical versus nonphysical may be misleading while discussing problems in quantum mechanics. Thus Landauer's analysis is also theoretically correct for mesoscopic systems that are made of semiconductors like GaAs/AlGaAs in which electron propagation is that of free electrons with an effective mass.

1.2 Aharonov-Bohm Effect

A charged quantum particle in a magnetic field can exhibit this effect. Schrodinger Eq. for a particle (spinless) is

$$i\hbar\frac{\partial\psi}{\partial t} = H\psi \tag{1.12}$$

The Hamiltonian for the particle with charge e_0 in magnetic field will be

$$H = \frac{1}{2m_0}(\hat{p} - \frac{e_0}{c}\vec{A})^2 + e_0 U \tag{1.13}$$

$\vec{A}$ is the vector potential, U is the scalar potential, and $\hat{p} = \frac{\hbar}{i}\vec{\nabla}$. We know from Maxwell's equation that

$$\vec{B} = \vec{\nabla} \times \vec{A} \tag{1.14}$$

Gauge transformation is defined as

$$\vec{A} \rightarrow \vec{A} + \vec{\nabla}\Lambda = A' \tag{1.15}$$

$$U \rightarrow U - \frac{1}{c}\frac{\partial\Lambda}{\partial t} = U' \tag{1.16}$$

$$\psi \rightarrow W\psi = \psi' \tag{1.17}$$

$$W = e^{\frac{ie_0\Lambda}{\hbar c}} \tag{1.18}$$

The last two Eqs. 1.17 and 1.18 are necessary for quantum particles only and not for classical particles. Let us try to use the freedom provided by these transformations. The first of these in Eq. 1.15 is easy to see as $\vec{\nabla} \times \vec{\nabla}\Lambda = 0$ for any scalar Λ, which means $\vec{A}$ and $\vec{A}'$ both give the same magnetic field $\vec{B}$. Schrödinger equation from Eqs. 1.12 and 1.13 after transformation is

$$i\hbar\frac{\partial\psi'}{\partial t} = \Big[\frac{1}{2m_0}\Big(\frac{\hbar}{i}\vec{\nabla} - \frac{e_0}{c}\vec{A}'\Big)^2 + e_0 U'\Big]\psi' \tag{1.19}$$

Schrödinger equation from Eqs. 1.12 and 1.13 before transformation is

$$i\hbar\frac{\partial\psi}{\partial t} = \Big[\frac{1}{2m_0}\Big(\frac{\hbar}{i}\vec{\nabla} - \frac{e_0}{c}\vec{A}\Big)^2 + e_0 U\Big]\psi \tag{1.20}$$

The claim is that the potentials $\vec{A}$ and U, as well as the Schrödinger equation, are all equivalent before and after transformation, which means we may work with either Eq. 1.19 or Eq. 1.20, whichever is more convenient. Note that for an electron $e_0 < 0$ while for a positron $e_0 > 0$. The equivalence of the two equations can be seen as follows. From Eq. 1.13,

$$H = \frac{1}{2m_0}\Big(\hat{p}^2 - \frac{e_0}{c}\hat{p}\cdot\vec{A} - \frac{e_0}{c}\vec{A}\cdot\hat{p} + \frac{e_0^2}{c^2}\vec{A}^2\Big) + e_0 U \tag{1.21}$$

In quantum mechanics since $\hat{p}$ is a differential operator,

$$\hat{p}\cdot\vec{A} = \frac{\hbar}{i}(\vec{\nabla}\cdot\vec{A}) + \vec{A}\cdot\hat{p} \tag{1.22}$$

The equivalence of LHS and RHS is with respect to their actions on a wave-function. So

$$H = \frac{\hat{p}^2}{2m_0} + \frac{i\hbar e_0}{2m_0 c}(\vec{\nabla}\cdot\vec{A}) - \frac{e_0}{m_0 c}\vec{A}\cdot\hat{p} + \frac{e_0^2}{2m_0 c^2}\vec{A}^2 + e_0 U \tag{1.23}$$

Note that we may use Coulomb gauge by which $\vec{\nabla}\cdot\vec{A} = 0$ but that can be inconvenient when generalizing to rings with width and thickness. It is better to use the following identity

$$\Big(\frac{\hbar}{i}\vec{\nabla} - \frac{e_0}{c}\vec{A}'\Big)W\psi = W\Big(\frac{\hbar}{i}\vec{\nabla} - \frac{e_0}{c}\vec{A}\Big)\psi \tag{1.24}$$

W is not a constant, but in the above identity it behaves like a constant except $\vec{A}' \to \vec{A}$. The LHS is called gauge covariant derivative. Let us prove the identity stated in Eq. 1.24.

$$LHS = \frac{\hbar}{i}(\vec{\nabla}W)\psi + W\frac{\hbar}{i}\vec{\nabla}\psi - \frac{e_0}{c}\vec{A}W\psi - \frac{e_0}{c}(\vec{\nabla}\Lambda)W\psi$$

Note that we have used Eq. 1.15. Substituting for $W = e^{\frac{ie_0\Lambda}{\hbar c}}$ from Eq. 1.18

$$\frac{\hbar}{i}\frac{ie_0}{\hbar c}(\vec{\nabla}\Lambda)W\psi + W\Big(\frac{\hbar}{i}\vec{\nabla} - \frac{e_0}{c}\vec{A}\Big)\psi - \frac{e_0}{c}(\vec{\nabla}\Lambda)W\psi = RHS$$

Using this identity one can show that Eqs. 1.19 and 1.20 are the same as shown below. We start from Eq. 1.19, i.e.,

$$i\hbar\frac{\partial\psi'}{\partial t} = \Big[\frac{1}{2m_0}\Big(\frac{\hbar}{i}\vec{\nabla} - \frac{e_0}{c}\vec{A}'\Big)^2 + e_0 U'\Big]\psi'$$

Using the identity in Eqs. 1.24 and 1.17

$$i\hbar\frac{\partial}{\partial t}(W\psi) = \frac{1}{2m_0}\Big(\frac{\hbar}{i}\vec{\nabla} - \frac{e_0}{c}\vec{A}'\Big)W\Big(\frac{\hbar}{i}\vec{\nabla} - \frac{e_0}{c}\vec{A}\Big)\psi + e_0 U' W\psi$$

or

$$i\hbar\frac{\partial}{\partial t}(W\psi) = \frac{1}{2m_0}\left(\frac{\hbar}{i}\vec{\nabla} - \frac{e_0}{c}\vec{A}'\right)W\psi'' + e_0U'W\psi$$

where

$$\left(\frac{\hbar}{i}\vec{\nabla} - \frac{e_0}{c}\vec{A}\right)\psi = \psi'' \tag{1.25}$$

or

$$i\hbar\frac{\partial}{\partial t}(W\psi) = \frac{1}{2m_0}W\left(\frac{\hbar}{i}\vec{\nabla} - \frac{e_0}{c}\vec{A}\right)\psi'' + e_0U'W\psi$$

or

$$i\hbar\frac{\partial\psi}{\partial t} = \frac{1}{2m_0}\left(\frac{\hbar}{i}\vec{\nabla} - \frac{e_0}{c}\vec{A}\right)\psi'' + e_0U'\psi$$

as W is independent of t. Thus using Eq. 1.25 and noting that in our case $U' = U$ (independent of t), we get

$$i\hbar\frac{\partial\psi}{\partial t} = \frac{1}{2m_0}\left(\frac{\hbar}{i}\vec{\nabla} - \frac{e_0}{c}\vec{A}\right)^2\psi + e_0U\psi \tag{1.26}$$

Thus starting from Eq. 1.19 we have recovered Eq. 1.20 and so we can solve any one of the two equations depending on convenience. It can be noted that in Eq. 1.19 we have the freedom to set

$$\vec{A}' = \vec{A} + \vec{\nabla}\Lambda = 0 \tag{1.27}$$

as Λ can be chosen arbitrarily, which gives

$$i\hbar\frac{\partial\psi'}{\partial t} = \left[\frac{1}{2m_0}\left(\frac{\hbar}{i}\vec{\nabla}\right)^2 + e_0U'\right]\psi' \tag{1.28}$$

With $U' = U$, this is exactly the Sc. Eq. without the magnetic field but ψ' has an extra phase as can be seen from Eqs. 1.17 and 1.18 that accounts for the magnetic field in the center of the ring. Since there is no magnetic field in the ring, there is no Lorentz force acting on the electrons. Now the question is whether an extra phase in the complex wave-function at all affect observables, specially when the probability is $|\psi'|^2$? The claim is that when closed trajectories are possible in the system then it will, and so this is a purely quantum interference phenomenon. This will result in an equilibrium current in the system that is not driven by any classical force. If the system does not allow any closed trajectories then there will be no observable effect. There is no restriction on the nature of the closed trajectory. It could be circular, or square, or elliptical or random walk that eventually comes back to the starting point. Of course if we take a perfectly circular 1D ring then there is only the option to move circularly. But as soon as we want to generalize to a ring of finite thickness then one can choose from various types of such closed trajectories. So it is important to understand it from the topological point of view that we only need the possibility of trajectories that close on itself.

Gauge fixing: Normally to solve the Sc. Eq. in presence of magnetic field one does a gauge fixing. According to Eqs. 1.14 and 1.15, Λ can be anything but we have already used this freedom in Eq. 1.27 that involves a derivative. So the components of $\vec{\nabla}\Lambda$ have to be fixed which is known as gauge fixing. By gauge fixing we make a particular choice of A_x, A_y, and A_z, such that the contour integral of the phase picked up in going round the flux turns out to be definite and independent of the details of the path. There can be many different choices for the gauge fixing and none of the popular choices is appropriate for our problem. Let us demonstrate this for one such popular choice for gauge fixing called the Coulomb gauge, i.e., $\vec{\nabla} \cdot \vec{A} = 0$. Of course we can write this down in polar form to simplify our calculations if our ring is perfectly circular and lying on a 2D plane. On the other hand if we can restrict such that an electron has only the freedom to move in a square path of sides of length "a", then it may be more suitable to choose the Coulomb gauge in Cartesian coordinates. The Coulomb gauge in Cartesian coordinates will be $\vec{\nabla}.\vec{A} = \partial_x A_x + \partial_y A_y + \partial_z A_z = 0$. That leads to

$$A_x = -B_0 y, \ A_y = 0, \ A_z = 0 \tag{1.29}$$

That in turn leads to a magnetic field that can be applied in a laboratory

$$\vec{B} = \vec{\nabla} \times \vec{A} = \hat{k} B_0$$

as one can see from Eq. 1.29 that

$$B_z = \partial_x A_y - \partial_y A_x = B_0$$

$$B_x = \partial_y A_z - \partial_z A_y = 0$$

$$B_y = \partial_z A_x - \partial_x A_z = 0$$

Thus Coulomb gauge in Cartesian coordinates gives rise to the vector potential profile depicted in Fig. 1.2. However, our choice of coordinate system should not determine the outcome of an observation and so one should be able to solve the situation of a perfectly circular ring with the kind of vector potential distribution shown in Fig. 1.2 and get the same result as that of gauge fixing in polar coordinates. So gauge fixing may not be at all a good idea in case the ring has a finite thickness or for a disk because in such systems one can imagine all kinds of closed trajectories ranging from circular to square.

Twisted periodic boundary condition: So one treats the problem using what is known in literature as the twisted periodic boundary condition according to which a closed trajectory in a finite thickness ring means $\psi(x, y, z) = \psi(x + L_x, y, z)$. The analysis works for any arbitrary closed trajectory that one can imagine in a finite thickness ring (a disk also allows such closed trajectories as a finite thickness ring is just a punctured disk). These trajectories are not directly related to electron motion

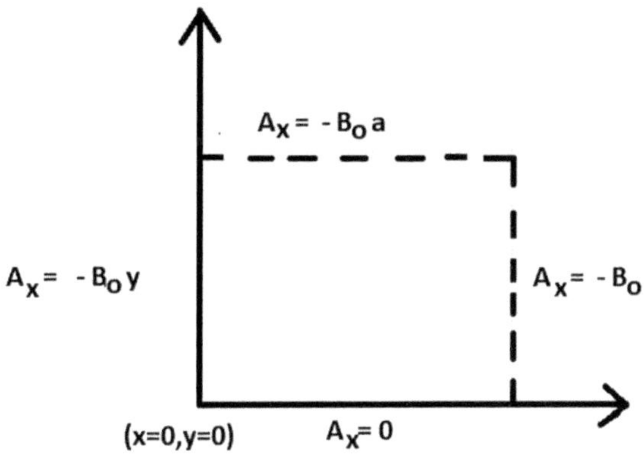

Fig. 1.2 Vector potential profile due to Coulomb gauge in Cartesian coordinates along a square path of sides of length "a" as given by Eq. 1.29

in the system but effectively related through the concept of Feynman paths that will be further elaborated in Sect. 1.5. For the special case of a 1D ring one can just drop the y and z degrees of freedom.

From Eqs. 1.17 and 1.18

$$\psi'(x, y, z) = e^{-\frac{ie_0}{\hbar c}\Lambda}\psi(x, y, z) \tag{1.30}$$

$$\psi'(x + L_x, y, z) = e^{-\frac{ie_0}{\hbar c}\int_0^{L_x}\vec{\nabla}\Lambda\cdot\vec{dl}}\psi(x + L_x, y, z) = e^{-\frac{ie_0}{\hbar c}\int_0^{L_x}\vec{\nabla}\Lambda\cdot\vec{dl}}\psi(x, y, z) \tag{1.31}$$

because a closed trajectory implies $\psi(x + L_x, y, z) = \psi(x, y, z)$ for ψ obtained from Eq. 1.20, while, ψ' satisfies Eq. 1.17. Here $\vec{dl}$ is an infinitesimal change in length along a given trajectory that make a closed path. Total length of the path being L_x. For the special situation of $L_x = 0$ we must get

$$\psi'(x, y, z) = e^{-\frac{ie_0}{\hbar c}\int_0^{0}\vec{\nabla}\Lambda\cdot\vec{dl}}\psi(x, y, z) = \psi(x, y, z) \tag{1.32}$$

Therefore, from Eqs. 1.31 and 1.32

$$\psi'(x + L_x, y, z) = e^{-\frac{ie_0}{\hbar c}\int_0^{L_x}\vec{\nabla}\Lambda\cdot\vec{dl}}\psi'(x, y, z) \tag{1.33}$$

Using Eq. 1.27, $-\vec{\nabla}\Lambda = \vec{A}$

$$\psi'(x + L_x, y, z) = e^{\frac{ie_0}{\hbar c}\int_0^{L_x}\vec{A}\cdot\vec{dl}}\psi'(x, y, z) \tag{1.34}$$

or

$$\psi'(x + L_x, y, z) = e^{\frac{i2\pi\Phi}{\Phi_0}}\psi'(x, y, z) \tag{1.35}$$

$\Phi = \int_0^{L_x} \vec{A} \cdot \vec{dl}$ is the flux through the ring and $\Phi_0 = \frac{hc}{e_0}$ is the flux quantum. Note that the only aspect of the ring geometry that has gone into this analysis is the fact that $x + L_x = x$ which means it can be an ellipse or any deformed version of a ring. To be mathematically correct the analysis is for the topology of a ring and not ring geometry.

In the rest of the chapter we will use the simple system of a one-dimensional ring because that can be solved analytically and the results can be elaborated with respect to the underlying physics. The magnetic field that affects the phase of the electron wave-function in the ring will magnetize the ring as we will show subsequently. This is different from all classical magnetizations and a few Gauss of magnetic field can magnetize the ring and is known as the celebrated Aharonov-Bohm effect. Realistic systems like a ring of finite thickness or disk will also get magnetized by a fraction of a single flux quantum based on the same physical principle. For such systems solving and demonstrating explicitly may involve some technical difficulties.

1.3 An Open Aharonov-Bohm Ring

Now consider the system schematically shown in Fig. 1.3 that was first studied by Buttiker [5] and help us understand the Landauer-Buttiker formalism for mesoscopic systems in a simple way. This is not an isolated ring but a ring connected to an electron reservoir making it a grand canonical system or an open system. Landauer's derivation of Eq. 1.11 was obtained by taking two such reservoirs one on each side of Fig. 1.1 and a non-equilibrium situation can be created by taking a chemical potential difference or a temperature difference between them. The temperature difference between them results in the different Fermi functions for electrons coming from the

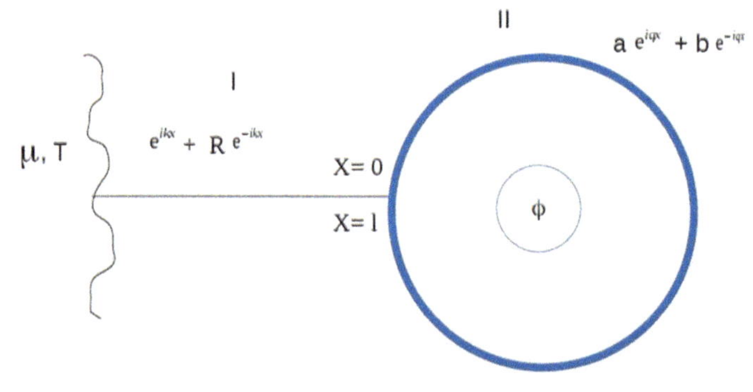

Fig. 1.3 A typical mesoscopic grand canonical system. It consists of an open 1D ring pierced by an Aharonov-Bohm flux Φ. The region inside the ring, shown by the thick line, has a constant potential U. A 1D thin solid line connects the ring to a reservoir and it is called lead. We have taken potential in the lead to be zero. Chemical potential and temperature of the reservoir are μ and T, respectively

left and those coming from the right. Landauer calls the reservoir on the left to be a perfect source of electrons and that on the right as a perfect sink. The ring is referred to as the mesoscopic system that may be made of semiconductor or metal. It can be literally made 1D or decoupled into multiple 1D channels [1, pp. 59] by lateral confinement that will be discussed in Chap. 4. The single reservoir in Fig. 1.3 is shown to have a chemical potential μ and temperature T. With one reservoir we cannot have a chemical potential difference or temperature difference and hence this is a system in equilibrium. The system has quantum states while the reservoir has electrons. The 1D thin line connecting the system and the reservoir is known as the lead. It can be made of the same material as the ring. The lead allows the electrons in the reservoir to populate the states in the ring and also allow the states in the ring to leak out of the ring. The electrons in the reservoir are completely thermalized classical particles. Total length of the lead and the circumference of the ring is less than the single particle coherence length of the quantum mechanical electrons at the temperatures in which mesoscopic experiments are conducted. So the electron dynamics in the lead and in the ring is governed by quantum mechanics. The single particle Schrodinger equation with an effective mass works very well for most of these systems. So we can write down a typical Fourier component of a wave-packet, which is also the solution of the 1D Schrodinger equation (as shown in Fig. 1.3) in the lead and the ring and match them at the junction between the ring and the lead. No wave-function is written inside the reservoir and the wave-function of an electron coming from the reservoir is taken as ae^{ikx} where x represents the coordinate and k the wave vector. The amplitude a is set to 1 as has been described in Eq. 1.5. Any initial phase in a do not affect the boundary condition equations in Sect. 1.4. This picture as envisaged by Landauer, therefore, claims a sharp line of demarcation between the quantum system of the lead and ring with the classical reservoir. While the axiomatic approach presented in Sect. 1.1 and explicitly stated in Eq. 1.4 does not require such a sharp line of demarcation. Axiomatic approaches lead to an objective theory, while Landauer's claim manifests. The sharp line of demarcation between a classical reservoir and a quantum system results in a contact resistance of 12.9 kilo ohms per sub-band that can be measured in multiple manifestations [1, pp. 53]. What manifests in knowledge (sense or gyan) is an objective reality. An isolated ring will have eigen-energies and eigen-states that will be explained in more detail in Chap. 2. In this picture of Fig. 1.3, the eigen-states of the ring can leak out resulting in level broadening and life time related effects which very accurately models dissipation and decoherence of realistic mesoscopic systems. It is this broadening effect that is captured quantitatively by scattering theory while all approaches starting from Hamiltonian try to parametrize this broadening in terms of a self energy. The non-equilibrium Green's function method discussed in the book by S. Datta is very suited to do this parametrization. However, many physical phenomenon may be overlooked by this parametrization and it is the scattering problem solved quantum mechanically that correctly incorporates evanescent modes and the presence of Fano resonances due to bound states in the continuum.

We believe that this book will emerge as a textbook on mesoscopic physics for physics students owing to the reasons given below. In statistical mechanics we deal with large systems that are connected to an electron reservoir. We start by saying that the total Hamiltonian of the system and reservoir can be decoupled as $H = H_s + H_r$ where H_s is the system Hamiltonian that gives some single particle or many body eigen-energies ϵ_s. The degrees of freedom in H_r, the Hamiltonian of the reservoir, can be integrated out to lead to parameters like chemical potential μ and temperature T. Now from ϵ_s, μ, and T one can derive a free energy that determines transport and thermodynamic properties. This will obviously fail as the system size starts becoming smaller and mesoscopic physics helps us understand step by step how it happens. First of all statistical mechanics ignores a surface term in the Hamiltonian which is of the form say H_{sr} that depends on the combined degrees of freedom of the system and that of the reservoir. In mesoscopic physics the leads take care of such terms and this is the first reason why we need the leads. At one end it is connected to the ring or the sample and reservoir at the other end. Secondly, the Hamiltonian will only give the electronic states and so where are the electrons? Statistical mechanics tell us that we do not have to be very particular about that. The electrons are anyway there and the only thing that matters is their fluctuations for which the law of large numbers applies (Stirling formula, etc.). All these ideas break down for mesoscopic systems in the laboratory. Law of large numbers has to be discarded and ensemble averaging does not work. Exchange of electrons takes place between the reservoirs and the system via leads that model the voltage probes, current probes, or any other probe like a scanning tunneling microscope tip, etc. We have to consider their specific roles and configurations through the electronic Hamiltonian in the leads. The role of leads can be partially included with the help of a self energy term in H_s that can account for level broadening as the energy levels of H_s leak out. One has to solve the scattering problem to account for this and thereby understand the effective changes in the eigen-energies and states of the system. These changes cannot be always modeled through parameters as is done via non-equilibrium Green's function techniques [1] because these changes are responsible for mesoscopic phenomena involving resistance, capacitance, and inductance. These changes can be explicitly calculated by solving the scattering problem and to set up the scattering problem we need the lead is the third reason. This will further give us concepts like partial states and local states that cannot be defined from the Hamiltonian alone. As a fourth reason, something very drastic happens because of the leads for which so far there is no way to account for by correcting eigen-energies of the system using some parameters. A state in the system can leak out through the leads in multiple ways, some of them being evanescent modes. This leads to bound states in the continuum and very dramatic changes in properties of the system. Without considering their effect mesoscopic systems cannot be properly understood and will be discussed in Chaps. 4, 5, and 6.

The two reservoirs in Fig. 1.1 (not shown) model the two battery terminals that drive the current through the system. Voltage probes (mesoscopic analogues of a voltmeter) and current probes (mesoscopic analogues of ammeters) also effectively act as additional electron reservoirs. Each has its own chemical potential and temperature and each can induce dephasing and decoherence [1]. It makes a lot of sense to study mesoscopic systems coupled to an arbitrary number of reservoirs including one and two and all of them correspond to some physical situation that is experimentally relevant. At the temperatures at which mesoscopic experiments are done, the experiments are mostly understood in terms of zero temperature calculations. So we will all along try to understand the zero temperature limit that we get from quantum mechanics. Also many body effects will not be considered in this book essentially because that has not developed to a theory that can be presented as textbook material. Many body effects have been mostly studied through numerical techniques. However, according to Eq. 1.4 we cannot write down any cross terms in the leads, so we cannot write a Slater determinant in the leads. But we can definitely write a Slater determinant in the sample (which in this case is the ring). So electron-electron interactions will give a correlated environment in the sample, with states in the leads being free electron states. Reference [4] shows that the main formalism discussed here, that of the Larmor clock, allows us to include many body effects in such a situation. We will now proceed to show that the system described in Fig. 1.3 will carry an equilibrium current called persistent current that does not require a power source. It is a purely mesoscopic phenomenon and very ideal to get insights into mesoscopic physics. This current magnetizes the ring and has been observed in rings made of gold or semiconductors [6]. Details of the results can be seen in the 1985 paper by Buttiker. Here we will provide an exact and hence easy to understand quantum mechanical treatment of the problem as compared to the adhoc theoretical treatment in the original paper by Buttiker. The reservoir injects electrons into the lead which is denoted as region I in the figure. These electrons populate the states inside the ring which is denoted as region II in the figure. The lead and the ring are connected at a point say J_1 (J_1 not marked in figure as it is easy to see the junction). The equation of motion in region I (or lead) and region II (or ring or system) is therefore the Schrodinger equation and the wave-functions in the two regions in absence of magnetic field can be written as

$$\psi_I = \frac{1}{\sqrt{k}}(e^{ikx} + Re^{-ikx}) \tag{1.36}$$

$$\psi_{II} = \frac{1}{\sqrt{q}}(ae^{iqx} + be^{-iqx}) \tag{1.37}$$

Here

$$q = \left(\frac{2m_0}{\hbar^2}(E-U)\right)^{1/2} \text{ and } k = \left(\frac{2m_0}{\hbar^2}E\right)^{1/2} \tag{1.38}$$

Note the notation a, b is not to be confused with that in Fig. 1.1 and for the rest of this chapter the definition of a and b will remain the same. The set up therefore, as envisaged by Landauer, has a sharp line of divide between the classical reservoir and the quantum lead and this is now well understood to correspond to the reality when the total length of the lead and the ring is less than the inelastic scattering length of the electron. Consistency checks can be made with other formalisms like Kubo formula or Boltzmann transport or Keldysh formalism, etc. [1]. One can bypass all the literature and research leading to it by adopting the simplified picture given in

Sect. 1.1, that is one may think that the lead continues inside the reservoir and as the electron goes far from the ring it reacts with the classical environment inside the reservoir that randomizes its phase.

The normalization constants used in Eqs. 1.36 and 1.37 are consistent with Eq. 1.10. One can either solve by removing the normalization constant of $\frac{1}{\sqrt{k}}$ and $\frac{1}{\sqrt{q}}$ by taking the incident electron is e^{ikx}, in which case one has to say that the DOS of the 1D lead is $\frac{1}{hv}$ which is the same as the DOS of an infinite 1D system. Or we can solve with this normalization constants which makes it easy to analytically continue the problem to include evanescent modes. At zero temperature counting states is the same as counting electrons. A typical example will be provided in Chap. 7, and one can see the discussions on Eqs. 7.27, 7.28, and 7.29. Note that 1D scattering problems in quantum mechanics textbooks will not approve of this normalization constant because they consider the 1D problem as an idealization that can be extended to 3D scattering which require a normalization constant of $\frac{1}{r}$ as the flux of spherically scattered plane wave will decrease as $\frac{3}{4\pi r^2}$. We will ignore the practice in textbooks for three reasons. First is that the Landauer-Buttiker formalism manifests. Secondly, one can position the normalization constant of Eqs. 1.36 and 1.37 with respect to the axioms of quantum mechanics and statistical mechanics. We will rather follow the view presented in the lecture series by F. Schuller in YouTube by which one can axiomatically establish a propagating wave-packet in quantum mechanics. This wave-packet is a propagating localized object in physical space and time. $a(k)e^{ikx}$ is a component or wavelet in this wave-packet and not the solution of the free Schrodinger equation. We can multiply any constant factor with $a(k)$ (the number line being affine) and try to find a differential equation that $a(k)e^{ikx}$ satisfies. Schuller establishes this equation in nearly his last lecture and it turns out to be the Schrodinger equation. Now this entire series of lectures does not depend on whether we take $a(k) = 1$ or $a(k)$ given by Eq. 1.10 or $a(k) \rightarrow \frac{a'(k)}{\sqrt{k}}$. We can use $\frac{1}{\sqrt{k}}$ to account for the DOS (or number of electrons) and $a(k)$ to determine the shape of the wave-packet. To get the correct evanescent mode current, we will see that one has to analytically continue $\frac{1}{\sqrt{k}}e^{ikx}$ and $\frac{1}{\sqrt{q}}e^{iqx}$. Explicit examples will be given in Chap. 7. If we use this freedom then we do not have to justify that the DOS in the semi-finite 1D lead is the same as that of an infinite 1D lead and box normalization like consolations can be done away with. Box normalization is a memorizing rule rather than a mathematical principle, that says apply it only to a square well potential (not a harmonic well potential for example) and therefore probability of finding the particle over entire space can be one at the cost of making the probability at a particular point go to zero, and we can switch between these two positions as per our convenience. Textbooks in quantum mechanics that propose $\frac{1}{r}$ as a normalization constant should consider the fact that $e^{i\vec{k}.\vec{r}}$ or e^{ikr} do not lead to the 3D Schrodinger equation but $\frac{1}{r}e^{ikr}$ does, which completely belittles the fact that normalization constant is a freedom. We believe that this $\frac{1}{r}$ normalization constant can be justified from the theory of quantum mechanics in curved surfaces that allows the use of polar coordinates but that is beyond the scope of this book. Thirdly, this is a standard practice in mesoscopic physics without much justification as can be seen in the references of Chap. 4 as well as in the definition of Green's function in mesoscopic systems [1, Eq. 3.3.6]. Box normalization like schemes rely on the fact that one can start from discrete countable states and then vary some parameter to bring them close and in the limit get the continuum. This is a mathematically wrong recipe to go from countable states to the uncountable continuum of states. The correct procedure is to follow a measure theory to quantify the volume of continuum of states making a wave-packet and that yields the same result as the 1D DOS. Nevertheless we do have to deal with integrations of the type $\int_{-\infty}^{\infty}|e^{ikx}|^2\mu(dx)$ to define an inner product in Hilbert space that is done with the help of rigged Hilbert space, Lebesgue integration, and equivalence relations, that is not directly related to normalization. We outline below how we can calculate some observable quantities by solving the scattering problem described in Fig. 1.3.

1.4 Boundary Conditions: Basic Introduction

One can find the equilibrium current flowing in the open system of Fig. 1.3 by solving a scattering problem that can be obtained from the boundary conditions. One can derive the current conserving boundary conditions starting from the time-independent Sc. Eq. corresponding to Eq. 1.28 where $U' = U$ (as there is no time dependence), $e_0 = 1$, and ψ' is written as ψ for simplicity.

$$-\frac{\hbar^2}{2m_0}\nabla^2\psi(\vec{r}) + U(\vec{r})\psi(\vec{r}) = E\psi(\vec{r}) \tag{1.39}$$

Integrating both sides over a volume V around the point where the lead meets the ring, we get

$$-\frac{\hbar^2}{2m_0}\int_V \nabla^2\psi(\vec{r})dV + \int_V U(\vec{r})\psi(\vec{r})dV = \int_V E\psi(\vec{r})dV \tag{1.40}$$

$$-\frac{\hbar^2}{2m_0}\int_V \vec{\nabla}.(\vec{\nabla}\psi(\vec{r}))dV + \int_V U(\vec{r})\psi(\vec{r})dV = \int_V E\psi(\vec{r})dV \tag{1.41}$$

Let $\vec{S}$ be the surface enclosing the volume V. In our 1D situation this volume V will be therefore three 1D lines meeting at a point and the surface **S** will be the three end points of the figure shown in Fig. 1.4.

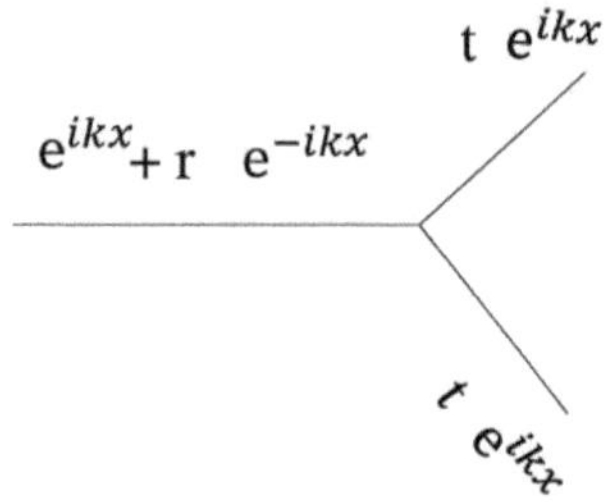

Fig. 1.4 The junction in Fig. 1.3 where the lead connects to the ring, in isolation

$$-\frac{\hbar^2}{2m_0}\lim_{V\to 0}\int_S (\vec{\nabla}\psi(\vec{r})).d\vec{S}+\lim_{V\to 0}\int_V U(\vec{r})\psi(\vec{r})dV=\lim_{V\to 0}\int_V E\psi(\vec{r})dV$$

We will use two conventions to denote a vector. One is by using an arrow head on top and the other is in bold font. Now as V is shrinking to 0, it follows that

$$\int_S (\vec{\nabla}\psi(\vec{r})).d\vec{S}=0$$

For the partial waves shown in Fig. 1.4 the integration becomes a sum and we can write

$$\sum \vec{\nabla}\psi(\vec{r})=0 \tag{1.42}$$

This is Kirchoff's law. Hence in one dimension, Eq. (1.42) can be written as

$$\sum_i \frac{d\psi_i}{dx_i}=0 \tag{1.43}$$

When this three prong structure of Fig. 1.4 is seen in isolation then the wave-functions obtained from Sc. Eq. are shown in Fig. 1.4. In that case one can apply these boundary conditions to get $r=-1/3$ and $t=2/3$. While when it becomes a part of the system in Fig. 1.3 then one has to use the wave-functions depicted in Fig. 1.3. R will depend on r and t and the details can be found below. Single valuedness of wave-function at the junction of Fig. 1.3 would imply

$$\frac{1}{\sqrt{k}}\Big(1+R\Big)=\frac{1}{\sqrt{q}}\Big(a+be^{-i\alpha}\Big)=\frac{1}{\sqrt{q}}\Big(ae^{iql+i\alpha}+be^{-iql}\Big) \tag{1.44}$$

And Eq. 1.43 implies

$$\Big[\frac{ik}{\sqrt{k}}(1-R)\Big]+\Big[\frac{iq}{\sqrt{q}}(-a+be^{-i\alpha})\Big]+\frac{iq}{\sqrt{q}}\Big[(ae^{iql+i\alpha}-be^{-iql})\Big]=0 \tag{1.45}$$

The first square bracket in Eq. 1.45 gives $\frac{d\psi}{dx}|_{x=0^-}$ which is a point in the lead close to the junction. One end of the ring starting from the junction is assigned a coordinate $x = 0$, while the other end is $x = l$. Second square bracket shows $\frac{d\psi}{dx}|_{x=0^+}$. And third square bracket gives $\frac{d\psi}{dx}|_{x=l^-}$. Thus Eq. 1.45 is obtained by writing Eq. 1.43 explicitly. Solving these three equations we can evaluate the three unknowns a, b, and R, and observables like magnetization will depend on them as we show below. The wave-functions shown in Eqs. 1.36 and 1.37 are in absence of magnetic field. Equations 1.44 and 1.45 are obtained from these wave-functions including the effect of the Aharonov-Bohm phase $\alpha = \frac{2\pi\Phi}{\Phi_0}$, where $\Phi_0 = \frac{hc}{e_0}$ is the fundamental flux quantum. However, only some terms acquire this phase and others do not. We will explain this in Sect. 1.5.

Although the current in the ring in absence of flux will be $|a|^2 - |b|^2$ and will obviously be zero, one cannot assume the same expression in presence of flux. In absence of flux a is the amplitude of clockwise moving electrons and b is the amplitude of anticlockwise electrons and they can only differ by a relative phase because none of them is preferred over the other. We give below a scheme to obtain the appropriate expression for current in presence of flux, which can be extended to other complicated situations as well, for example when we have evanescent modes inside the ring, or when some modes in the ring are propagating while some are evanescent, etc. Such multichannel situations have been solved numerically as they can have technological applications [7]. The 1D system considered here can be solved analytically and is a good example to understand realistic systems. Current conservation implies that sum of currents arriving and leaving the junction is 0. So we quantum mechanically calculate the current $\frac{e_0\hbar}{2m_0 i}[\psi^*\nabla\psi - HC]$ in region I and also at the two ends of region II. That for each region is written within square brackets and solved separately preserving current conservation in each step (for the time being assume the α dependence of the terms as in Eqs. 1.44 and 1.45 as their origin is explained in the next section).

$$\begin{aligned}
&\frac{1}{k}\left[(e^{ikx} + Re^{-ikx})^* \frac{d}{dx}(e^{ikx} + Re^{-ikx})\right]_{0^-} \\
&-\frac{1}{q}\left[(ae^{iqx} + be^{-iqx-i\alpha})^* \frac{d}{dx}(ae^{iqx} + be^{-iqx-i\alpha})\right]_{0^+} \\
&+\frac{1}{q}\left[(ae^{iqx+i\alpha} + be^{-iqx})^* \frac{d}{dx}(ae^{iqx+i\alpha} + be^{-iqx})\right]_{l^-} - HC = 0 \qquad (1.46)
\end{aligned}$$

Here HC stands for Hermitian conjugate. Prefactors like $\frac{e_0\hbar}{2m_0 i}$ can be added in the end. Clockwise current at $x = 0^+$ is leaving the junction and convention followed is currents leaving the junction are negative while currents approaching the junction are positive according to the gradient convention in Eq. 1.42. The first square bracketed term gives the current in the lead, that is immediate to the left of the junction in Fig. 1.3 or $x = 0^-$. The second square bracketed term gives the current at a point in the ring, that is denoted as $x = 0^+$ in the figure. The third square bracketed term gives the current at a point in the ring, that is denoted as $x = l^-$ in the figure. The equal sign follows from current conservation as there is no source in the junction. Simplifying we get

$$\left[\frac{ik}{k}(1+R^*)(1-R)\right] - \left[\frac{iq}{q}(a^* + b^* e^{i\alpha})(a - be^{-i\alpha})\right]$$
$$+ \left[\frac{iq}{q}(a^* e^{-iql-i\alpha} + b^* e^{iql})(ae^{iql+i\alpha} - be^{-iql})\right] - HC = 0 \tag{1.47}$$

or

$$\left[i - iR + iR^* - i \mid R \mid^2\right] - \left[i \mid a \mid^2 - ia^* be^{-i\alpha} + ib^* ae^{i\alpha} - i \mid b \mid^2\right]$$
$$+ \left[i \mid a \mid^2 - ia^* be^{-2iql-i\alpha} + ib^* ae^{2iql+i\alpha} - i \mid b \mid^2\right] - HC = 0 \tag{1.48}$$

The HC is written below

$$\left[-i + iR^* - iR + i \mid R \mid^2\right] - \left[-i \mid a \mid^2 + iab^* e^{i\alpha} - iba^* e^{-i\alpha} + i \mid b \mid^2\right]$$
$$+ \left[-i \mid a \mid^2 + iab^* e^{2iql+i\alpha} - iba^* e^{-2iql-i\alpha} + i \mid b \mid^2\right] \tag{1.49}$$

Subtracting Eq. 1.49 from Eq. 1.48 we get 0 using the condition $\mid R \mid^2 = 1$. It is expected on physical grounds that the net current in the lead will be 0 and the current at $x = 0^+$ will be equal to the current at $x = l^-$.

Let us consider only the second square bracket terms in Eqs. 1.48 and 1.49. That is

$$-i \mid a \mid^2 + ia^* be^{-i\alpha} - ib^* ae^{i\alpha} + i \mid b \mid^2$$

and

$$i \mid a \mid^2 - iab^* e^{i\alpha} + iba^* e^{-i\alpha} - i \mid b \mid^2$$

The difference between the two is

$$-2i \mid a \mid^2 + 2i \mid b \mid^2$$

which on including the $\frac{e_0 \hbar}{2m_0 i}$ gives the current inside the ring to be

$$\frac{e_0 \hbar}{m_0}(- \mid a \mid^2 + \mid b \mid^2) \tag{1.50}$$

This is always arbitrary to an overall sign for clockwise versus anticlockwise sign convention. So if $E > U$ then the current in the system or in the ring will be given by $\frac{e_0\hbar}{m_0}(|a|^2 - |b|^2)$. It can be easily calculated to check that if Φ is zero then there is no current in the system as is expected for an equilibrium situation. However for $\Phi \neq 0$ the current is finite. It is an equilibrium current called persistent current, purely quantum mechanical in origin and further explanation of it will be given in Chap. 2.

Here we would like to note that an interesting situation occurs when $E < U$ in Eq. 1.38. Then the persistent current is carried by tunneling electrons or evanescent modes. Quantum tunneling is still shrouded in mystery and we do know a few things about an electron that has tunneled through a barrier but we still know absolutely nothing about the state of the electron under the barrier. We will show in Chap. 7 that one can settle most of the controversy by studying persistent currents due to Aharonov-Bohm effect for the evanescent modes. In this situation $q \to is$ and the expression for the current can be derived using the same scheme as before, which means we substitute $q \longrightarrow is$ and from Eqs. 1.36 and 1.37, ψ_I and ψ_{II} become

$$\psi_I = \frac{1}{\sqrt{k}}\Big[e^{ikx} + Re^{-ikx}\Big] \tag{1.51}$$

$$\psi_{II} = \frac{1}{\sqrt{is}}\Big[ae^{i(is)x} + be^{-i(is)x}\Big] \tag{1.52}$$

Applying boundary conditions of continuity and current conservation (Eq. 1.43), analogues of Eqs. 1.44 and 1.45 become

$$\frac{1}{\sqrt{k}}(1+R) = \frac{1}{\sqrt{is}}(a + be^{-i\alpha}) = \frac{1}{\sqrt{is}}(ae^{i(is)l+i\alpha} + be^{-i(is)l}) \tag{1.53}$$

$$\Big[\frac{1}{\sqrt{k}}ik(1-R)\Big] + \Big[\frac{1}{\sqrt{is}}i(is)(-a + be^{-i\alpha})\Big] + \Big[\frac{1}{\sqrt{is}}i(is)(ae^{i(is)l+i\alpha} - be^{-i(is)l})\Big] = 0 \tag{1.54}$$

In Eq. 1.54, the first square bracket gives $\frac{d\psi_I}{dx}|_{x=0^-}$. The second square bracket shows $\frac{d\psi_{II}}{dx}|_{x=0^+}$. The third square bracket gives $\frac{d\psi_{II}}{dx}|_{x=l^-}$. So again we calculate for the wave-functions in Eqs. 1.51 and 1.52 with the usual α dependence

$$\frac{e_0\hbar}{2m_0 i}[\psi^*\nabla\psi - HC] \tag{1.55}$$

and analogue of Eq. 1.46 becomes

$$\frac{1}{k}\Big[(e^{ikx} + Re^{-ikx})^*\frac{d}{dx}(e^{ikx} + Re^{-ikx})\Big]_{0^-}$$

$$+\frac{1}{s}\Big[(ae^{i(is)x} + be^{-i(is)x-i\alpha})^*\frac{d}{dx}(ae^{i(is)x} + be^{-i(is)x-i\alpha})\Big]_{0^+}$$

$$-\frac{1}{s}\left[(ae^{i(is)x+i\alpha}+be^{-i(is)x})^*\frac{d}{dx}(ae^{i(is)x+i\alpha}+be^{-i(is)x})\right]_{l^-}-HC=0 \quad (1.56)$$

Evaluating the first three square brackets we get

$$\left[i-iR+iR^*-i\mid R\mid^2\right]-\left[\mid a\mid^2-a^*be^{-i\alpha}+b^*ae^{i\alpha}-\mid b\mid^2\right]+$$

$$\left[\mid a\mid^2 e^{-2sl}-a^*be^{-i\alpha}+b^*ae^{i\alpha}-\mid b\mid^2 e^{2sl}\right] \quad (1.57)$$

Considering the second square bracket term only for the current in the ring (the third square bracket can be alternately considered to give the same result) and substituting in the first term in Eq. 1.55 we get

$$-\frac{e_0\hbar}{2m_0 i}\left[\mid a\mid^2-a^*be^{-i\alpha}+b^*ae^{i\alpha}-\mid b\mid^2\right] \quad (1.58)$$

Now subtracting the Hermitian conjugate we get current expression in the ring

$$\frac{e_0\hbar}{m_0 i}\left[ba^*e^{-i\alpha}-ab^*e^{i\alpha}\right] \quad (1.59)$$

This will be further discussed in Chap. 7.

1.5 Inclusion of Magnetic Field: Feynman Path Approach

We have discussed in Sect. 1.2 that in a finite thickness ring or in a disk many alternate classical trajectories are possible for a quantum particle, in going from one point in the system to another point. A classical particle chooses one particular trajectory from them which is determined by the least action principle. However, a quantum particle does not do that and in fact there are several alternate physical pictures that different schools have introduced. All these alternate pictures lead to the same final result. For example the Copenhagen school says that the quantum particle propagates as a wave as long as we leave it alone but as soon as we try to observe it, it tries to acquire particle like behavior. The Feynman path approach is an alternate picture to that by which we say that the particle simultaneously go through all possible classical trajectories weighted down by the action along these trajectories that give a phase term. So different trajectories result in different phases and the amplitudes for different trajectories interfere to give the same final result as the Copenhagen school. Even for the 1D system of Fig. 1.3, an infinite number of trajectories are possible. Below we will use this Feynman path approach to justify the boundary conditions of Eqs. 1.44 and 1.45, which will also be a simple demonstration of how the Feynman

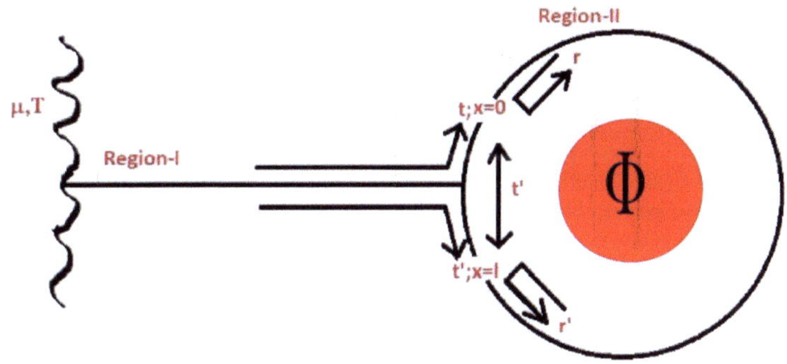

Fig. 1.5 Decomposition of the system in Fig. 1.3 into a three prong structure (depicted as region I) of Fig. 1.4 and a ring (depicted as region II). t and r are the transmission and reflection amplitudes of Fig. 1.4 which means $t = 2/3$ and $r = -1/3$. r' is related to r in the manner explained in Eq. 1.98. Time reversal symmetry implies $t' = t$

path approach is applied. This has been briefly touched upon in S. Datta's book on page 63, for the square well problem. Below we do it for the AB ring of Fig. 1.3.

In the Feynman path method, the full system is broken down into smaller parts as would be necessary and explained below. The system in Fig. 1.3 can be broken or decoupled into two parts. It is shown in Fig. 1.5 by breaking the lines at $x = 0^+$ and at $x = l^-$. Therefore the part depicted as region I is the three prong structure shown in Fig. 1.4. The part depicted as region II is the ring-like structure enclosing the flux. Feynman path method provides a good connection between canonical solutions (leading to Eqs. 1.44 and 1.45) in quantum mechanics and physical motion resulting in currents through a sample. In the following we present a calculation of the wave-function inside the ring at $x = 0^+$ and $x = l^-$, using the Feynman path approach. This will also justify the way in which the AB phase enters the boundary conditions expressed in Eqs. 1.44 and 1.45. The twist is that one has to sum the amplitudes for all possible classical trajectories (weighted by their classical action) starting at the near end of the lead ($x = 0^-$) and ending at the point of interest where we intend to find the wave-function ($x = 0^+$ or $x = l^-$) inside the ring. To systematically include all possible trajectories, we find a way to classify the trajectories and that can be done by the number of reflections involved and described in detail below as zero reflection terms, 1 reflection terms, etc.

Calculation of wave-function at $x = 0^+$: Zero reflection terms depicted and explained in Fig. 1.6

$$\begin{aligned}\psi_0^{tt}(0^+) &= t + te^{iql+i\alpha}t' + te^{iql+i\alpha}t'e^{iql+i\alpha}t' + \ldots \\ &= \frac{t}{1 - t'e^{iql+i\alpha}}\end{aligned} \tag{1.60}$$

Here superscript tt stands for "through top", subscript 0 stands for "zero reflection term", and 0^+ in parenthesis denotes the coordinate where we intend to evaluate

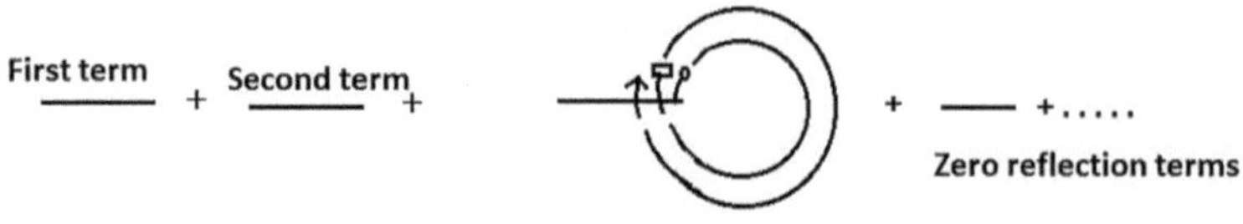

Fig. 1.6 The "First term" and the "Second term" are not drawn but can be figured out from the third term which is drawn explicitly. The third term corresponds to a Feynman trajectory where an electron comes from the reservoir and at the junction it is transmitted up to the position of the small circle. It then makes a full circle along the ring to reach the position of the small rectangle. And then it continues to make another full circle in the ring to get to the position where the arrow head points. As per the coordinate choice depicted in Fig. 1.3 or Fig. 1.5 this position is $x = 0^+$. The first term will be the situation if this third term is truncated at the small circle. The second term will be the situation when this third term is truncated at the small rectangle. The first term will give the first term in the series leading to Eq. 1.60. The second term will give the second term in the series in Eq. 1.60. The third term will give the third term in the series in Eq. 1.60. For example, getting to the small circle will involve a transmission amplitude of $t = 2/3$ (first term in Eq. 1.60). And then make a full circle propagation with appropriate action require us to multiply a factor $e^{iql+i\alpha}$ and another factor of t' (which may or may not be the same as t as will be discussed subsequently) to get to the small rectangle (second term in Eq. 1.60). And finally to get to the point $x = 0^+$ at the end of the arrow require another factor of $e^{iql+i\alpha}t'$ (third term). None of the terms in Eq. 1.60 depend on r or r' and so all the terms are classified as zero reflection terms

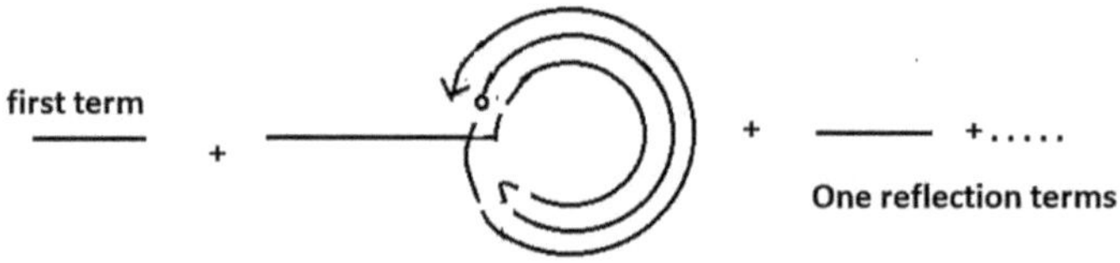

Fig. 1.7 One should be able to understand this figure by reading the caption of Fig. 1.6 and extend the logic. It depicts the Feynman paths leading to the series in Eq. 1.61

the wave-function. These symbols will become more evident as we proceed. One reflection terms corresponding to Fig. 1.7

$$\psi_1^{tt}(0^+) = te^{iql+i\alpha}r'e^{iql-i\alpha} + te^{iql+i\alpha}r'e^{iql-i\alpha}t'e^{iql-i\alpha} + \dots$$
$$= \frac{tr'e^{2iql}}{1 - t'e^{iql-i\alpha}} \tag{1.61}$$

Two reflection terms corresponding to Fig. 1.8

$$\psi_2^{tt}(0^+) = te^{iql+i\alpha}r'e^{iql-i\alpha}r + te^{iql+i\alpha}r'e^{iql-i\alpha}re^{iql+i\alpha}t' + \dots$$
$$= \frac{trr'e^{2iql}}{1 - t'e^{iql+i\alpha}} \tag{1.62}$$

Fig. 1.8 It depicts the Feynman paths leading to the series in Eq. 1.62

Three reflection terms (self explanatory from earlier discussion)

$$\psi_3^{tt}(0^+) = te^{iql+i\alpha}r'e^{iql-i\alpha}re^{iql+i\alpha}r'e^{iql-i\alpha} + ...t'e^{iql-i\alpha} + ...$$

$$= \frac{tr'^2re^{4iql}}{1-t'e^{iql-i\alpha}} \tag{1.63}$$

Four reflection terms (self explanatory from earlier discussion)

$$\psi_4^{tt}(0^+) = te^{iql+i\alpha}r'e^{iql-i\alpha}re^{iql+i\alpha}r'e^{iql-i\alpha}r + ...e^{iql+i\alpha}t' + ...$$

$$= \frac{tr^2r'^2e^{4iql}}{1-t'e^{iql+i\alpha}} \tag{1.64}$$

Five reflection terms (self explanatory from earlier discussion)

$$\psi_5^{tt}(0^+) = te^{iql+i\alpha}r'e^{iql-i\alpha}re^{iql+i\alpha}r'e^{iql-i\alpha}re^{iql+i\alpha}r'e^{iql-i\alpha} + ...t'e^{iql-i\alpha} + ...$$

$$= \frac{tr'^3r^2e^{6iql}}{1-t'e^{iql-i\alpha}} \tag{1.65}$$

Only five of the infinite number of terms are evaluated here to show that they form two geometric series. More details can be seen in the caption of Fig. 1.6. The even reflection terms in Eqs. 1.60, 1.62, and 1.64 form one series and the odd reflection terms in Eqs. 1.61, 1.63, and 1.65 form another.

So the contribution to the wave-function inside the ring at the point $x = 0^+$ coming from these diagrams in Figs. 1.6, 1.7, 1.8, etc., will be the sum of the RHS of Eqs. 1.60 to 1.65

$$\psi^{tt}(0^+) = [\psi_0^{tt}(0^+) + \psi_2^{tt}(0^+) + \psi_4^{tt}(0^+) + ...] + [\psi_1^{tt}(0^+) + \psi_3^{tt}(0^+) + \psi_5^{tt}(0^+) + ...]$$

$$\psi^{tt}(0^+) = \frac{t}{1-t'e^{iql+i\alpha}}(1 + rr'e^{2iql} + r^2r'^2e^{4iql} + ...)$$

$$+\frac{tr'e^{2iql}}{1-t'e^{iql-i\alpha}}(1 + r're^{2iql} + r^2r'^2e^{4iql} + ...)$$

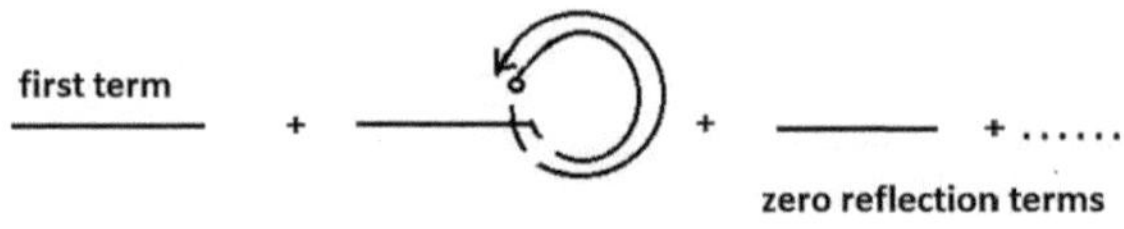

Fig. 1.9 It depicts the Feynman paths leading to the series in Eq. 1.67

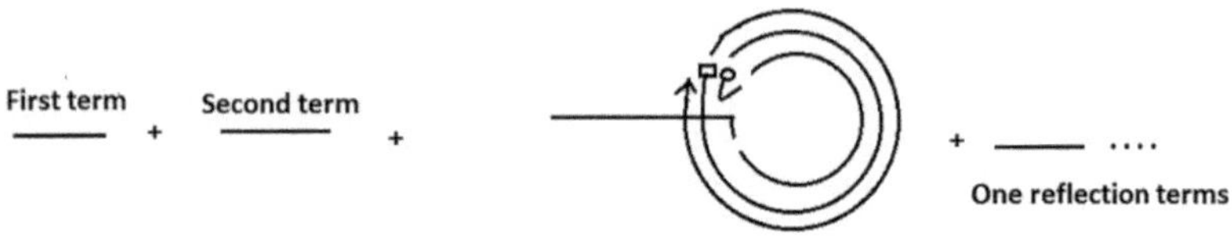

Fig. 1.10 It depicts the Feynman paths leading to the series in Eq. 1.68

or

$$\psi^{tt}(0^+) = \frac{t}{1 - t'e^{iql+i\alpha}} \frac{1}{1 - r'r e^{2iql}} + \frac{tr'e^{2iql}}{1 - t'e^{iql-i\alpha}} \frac{1}{1 - rr'e^{2iql}} \tag{1.66}$$

For the wave-function at $x = 0^+$, one can get a different set of Feynman paths and the difference can be seen by comparing Figs. 1.6 with 1.9. These terms too can be classified by the number of reflections involved. Zero reflection terms (see Fig. 1.9)

$$\psi_0^{tb}(0^+) = t'e^{iql-i\alpha} + t'e^{iql-i\alpha}t'e^{iql-i\alpha} + ...$$

$$= \frac{t'e^{iql-i\alpha}}{1 - t'e^{iql-i\alpha}} \tag{1.67}$$

tb now stands for "through bottom". One reflection terms (see Fig. 1.10)

$$\psi_1^{tb}(0^+) = t'e^{iql-i\alpha}r + t'e^{iql-i\alpha}re^{iql+i\alpha}t' + ...$$

$$= \frac{t'r e^{iql-i\alpha}}{1 - t'e^{iql+i\alpha}} \tag{1.68}$$

Two reflection terms

$$\psi_2^{tb}(0^+) = t'e^{iql-i\alpha}re^{iql+i\alpha}r'e^{iql-i\alpha} + ...t'e^{iql-i\alpha} + ...$$

$$= \frac{r'rt'e^{3iql-i\alpha}}{1 - t'e^{iql-i\alpha}} \tag{1.69}$$

Three reflection terms

$$\psi_3^{tb}(0^+) = t'e^{iql-i\alpha}re^{iql+i\alpha}r'e^{iql-i\alpha}r + ...e^{iql+i\alpha}t' + ...$$

$$= \frac{t'r'r^2 e^{3iql-i\alpha}}{1 - t'e^{iql+i\alpha}} \tag{1.70}$$

Four reflection terms

$$\psi_4^{tb}(0^+) = t'e^{iql-i\alpha} r e^{iql+i\alpha} r' e^{iql-i\alpha} r e^{iql+i\alpha} r' e^{iql-i\alpha} + \ldots.t'e^{iql-i\alpha} + \ldots$$

$$= \frac{t'r'^2 r^2 e^{5iql-i\alpha}}{1 - t'e^{iql-i\alpha}} \tag{1.71}$$

Five reflection terms

$$\psi_5^{tb}(0^+) = t'e^{iql-i\alpha} r e^{iql+i\alpha} r' e^{iql-i\alpha} r e^{iql+i\alpha} r' e^{iql-i\alpha} r + \ldots..e^{iql+i\alpha} t' + \ldots$$

$$= \frac{t'r^3 r'^2 e^{5iql-i\alpha}}{1 - t'e^{iql+i\alpha}} \tag{1.72}$$

So the contribution to the wave-function inside the ring at the point $x = 0^+$ coming from these diagrams will again form two geometric series,

$$\psi^{tb}(0^+) = [\psi_0^{tb}(0^+) + \psi_2^{tb}(0^+) + \psi_4^{tb}(0^+) + \ldots] + [\psi_1^{tb}(0^+) + \psi_3^{tb}(0^+) + \psi_5^{tb}(0^+) + \ldots]$$

$$\psi^{tb}(0^+) = \frac{t'e^{iql-i\alpha}}{1 - t'e^{iql-i\alpha}}(1 + r'r e^{2iql} + r'^2 r^2 e^{4iql} + \ldots..)$$

$$+ \frac{t'r e^{iql-i\alpha}}{1 - t'e^{iql+i\alpha}}(1 + r'r e^{2iql} + r^2 r'^2 e^{4iql} + \ldots)$$

$$\psi^{tb}(0^+) = \frac{t'e^{iql-i\alpha}}{1 - t'e^{iql-i\alpha}} \frac{1}{1 - r'r e^{2iql}} + \frac{t'r e^{iql-i\alpha}}{1 - t'e^{iql+i\alpha}} \frac{1}{1 - r'r e^{2iql}} \tag{1.73}$$

The net wave-function at $x = 0^+$

$$\psi(0^+) = \psi^{tt}(0^+) + \psi^{tb}(0^+)$$

$$= \frac{t}{1 - t'e^{iql+i\alpha}} \frac{1}{1 - rr'e^{2iql}} + \frac{tr'e^{2iql}}{1 - t'e^{iql-i\alpha}} \frac{1}{1 - rr'e^{2iql}}$$

$$+ \frac{t'e^{iql-i\alpha}}{1 - t'e^{iql-i\alpha}} \frac{1}{1 - rr'e^{2iql}} + \frac{t'r e^{iql-i\alpha}}{1 - t'e^{iql+i\alpha}} \frac{1}{1 - rr'e^{2iql}}$$

$$= \frac{t}{1 - rr'e^{2iql}} \left[\frac{1}{1 - t'e^{iql+i\alpha}} + \frac{r'e^{2iql}}{1 - t'e^{iql-i\alpha}} \right]$$

$$+\frac{t'e^{iql-i\alpha}}{1-rr'e^{2iql}}\left[\frac{1}{1-t'e^{iql-i\alpha}}+\frac{r}{1-t'e^{iql+i\alpha}}\right]$$

Thus it is of the form $a + be^{-i\alpha}$ where

$$a = \frac{t}{1-rr'e^{2iql}}\left[\frac{1}{1-t'e^{iql+i\alpha}}+\frac{r'e^{2iql}}{1-t'e^{iql-i\alpha}}\right] \tag{1.74}$$

$$b = \frac{t'e^{iql}}{1-rr'e^{2iql}}\left[\frac{1}{1-t'e^{iql-i\alpha}}+\frac{r}{1-t'e^{iql+i\alpha}}\right] \tag{1.75}$$

This explains the second term in Eq. 1.44 and the second bracketed term in Eq. 1.45 where the normalization constant is added. Also note that a and b as obtained above is completely known as t, t', r, and r' are known (see further Eq. 1.76 below). Thus plugging them in Eq. 1.50 we can directly get the current in the ring without having to solve the simultaneous equations in (1.44) and (1.45). Also note that

$$r'e^{2iql} = r \tag{1.76}$$

as we will explain below implying that in absence of flux there will be no current in the ring. And that also implies that the flux will break the symmetry between a and b to give rise to a current in the ring which will magnetize the ring. We would like to emphasize again that this is a purely equilibrium current due to quantum interference and Aharonov-Bohm effect. It does not require any classical force to sustain it and is a perpetual current. At this point one may dive into the vast amount of literature to see how the current flows in the ring but the heat is dissipated in the reservoir, implying a sharp spatial separation between the two processes, the first one being quantum and the second being classical. One may make the length of the lead tending to zero as that length never entered our analysis. Perpetual currents cannot survive in a dissipative ring. The length of the ring is less than the inelastic mean free path of the electron and these currents have been observed experimentally. However strange it may seem, it is this picture of a reservoir that is physical and will be extended to real 3D systems in Chap. 4 while the view presented in Sect. 1.1 is an idealization to the situation of 1D semi-infinite leads. Now the literature is so extensive that we are not making any attempt to refer such a well established phenomenon other than the book by S. Datta. The simple analysis presented here from first principles should make it clear.

Calculation of wave-function at $x = l^-$: In the same way we proceed to calculate the wave-function at $x = l^-$. Zero reflection terms (see Fig. 1.11)

$$\psi_0^{tb}(l^-) = t' + t'e^{iql-i\alpha}t' + t'e^{iql-i\alpha}t'e^{iql-i\alpha}t' + ...$$

$$= \frac{t'}{1-t'e^{iql-i\alpha}} \tag{1.77}$$

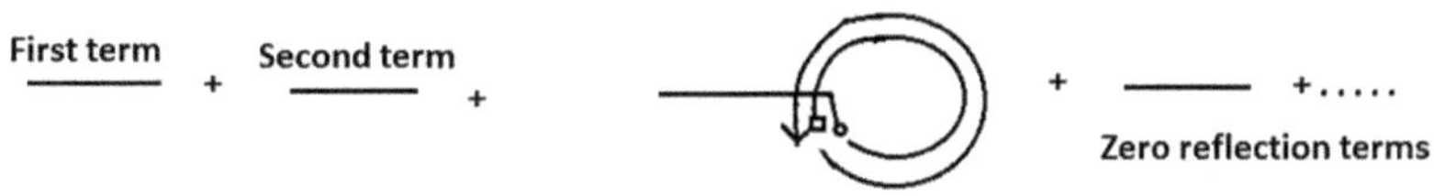

Fig. 1.11 It depicts the Feynman paths leading to the series in Eq. 1.77

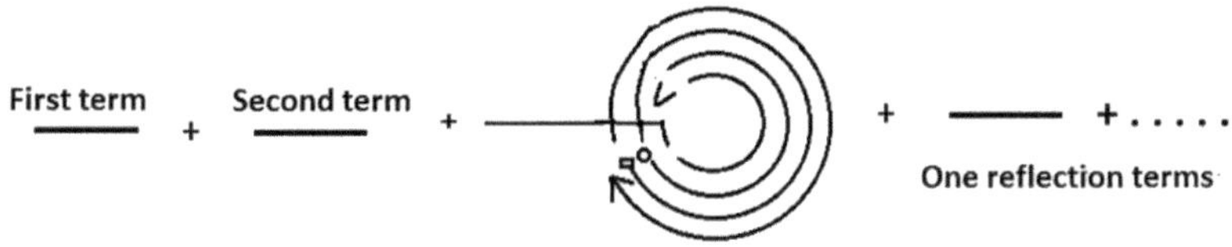

Fig. 1.12 It depicts the Feynman paths leading to the series in Eq. 1.78

One reflection terms (see Fig. 1.12)

$$\psi_1^{tb}(l^-) = t'e^{iql-i\alpha}re^{iql+i\alpha} + t'e^{iql-i\alpha}re^{iql+i\alpha}t'e^{iql+i\alpha} + ...$$

$$= \frac{t're^{2iql}}{1 - t'e^{iql+i\alpha}} \tag{1.78}$$

Two reflection terms

$$\psi_2^{tb}(l^-) = t'e^{iql-i\alpha}re^{iql+i\alpha}r' + t'e^{iql-i\alpha}re^{iql+i\alpha}r'e^{iql-i\alpha}t' + ...$$

$$= \frac{t'rr'e^{2iql}}{1 - t'e^{iql-i\alpha}} \tag{1.79}$$

Three reflection terms

$$\psi_3^{tb}(l^-) = t'e^{iql-i\alpha}re^{iql+i\alpha}r'e^{iql-i\alpha}re^{iql+i\alpha} + ...t'e^{iql+i\alpha} + ...$$

$$= \frac{t'r^2r'e^{4iql}}{1 - t'e^{iql+i\alpha}} \tag{1.80}$$

Four reflection terms

$$\psi_4^{tb}(l^-) = t'e^{iql-i\alpha}re^{iql+i\alpha}r'e^{iql-i\alpha}re^{iql+i\alpha}r' + ...e^{iql-i\alpha}t' + ...$$

$$= \frac{t'r^2r'^2e^{4iql}}{1 - t'e^{iql-i\alpha}} \tag{1.81}$$

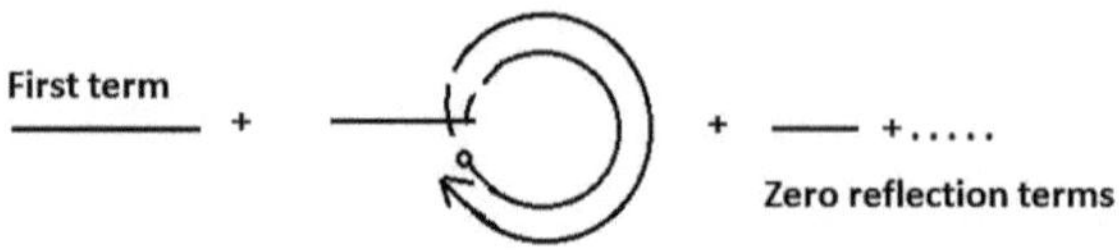

Fig. 1.13 It depicts the Feynman paths leading to the series in Eq. 1.84

Five reflection terms

$$\psi_5^{tb}(l^-) = t'e^{iql-i\alpha}re^{iql+i\alpha}r'e^{iql-i\alpha}re^{iql+i\alpha}r'e^{iql-i\alpha}re^{iql+i\alpha}$$
$$+.......t'e^{iql+i\alpha} + ...$$
$$= \frac{t'r^3r'^2e^{6iql}}{1-t'e^{iql+i\alpha}} \tag{1.82}$$

So the contribution to the wave-function inside the ring at the point $x = l^-$ coming from these diagrams will form two geometric series. One of them is given by the Eqs. 1.77, 1.79, 1.81, and so on and the other by the Eqs. 1.78, 1.80, 1.82, and so on to give

$$\psi^{tb}(l^-) = [\psi_0^{tb}(l^-) + \psi_2^{tb}(l^-) + \psi_4^{tb}(l^-) + ...] + [\psi_1^{tb}(l^-) + \psi_3^{tb}(l^-) + \psi_5^{tb}(l^-) + ...]$$

$$\psi^{tb}(l^-) = \frac{t'}{1-t'e^{iql-i\alpha}}(1 + rr'e^{2iql} + r^2r'^2e^{4iql} + ...)$$
$$+\frac{t're^{2iql}}{1-t'e^{iql+i\alpha}}(1 + rr'e^{2iql} + r^2r'^2e^{4iql} + ..)$$

$$\psi^{tb}(l^-) = \frac{t'}{1-t'e^{iql-i\alpha}}\frac{1}{1-rr'e^{2iql}} + \frac{t're^{2iql}}{1-t'e^{iql+i\alpha}}\frac{1}{1-rr'e^{2iql}} \tag{1.83}$$

As in the previous case there is another set of diagrams for the wave-function at $x = l^-$. Zero reflection terms (see Fig. 1.13)

$$\psi_0^{tt}(l^-) = te^{iql+i\alpha} + te^{iql+i\alpha}t'e^{iql+i\alpha} + ...$$
$$= \frac{te^{iql+i\alpha}}{1-t'e^{iql+i\alpha}} \tag{1.84}$$

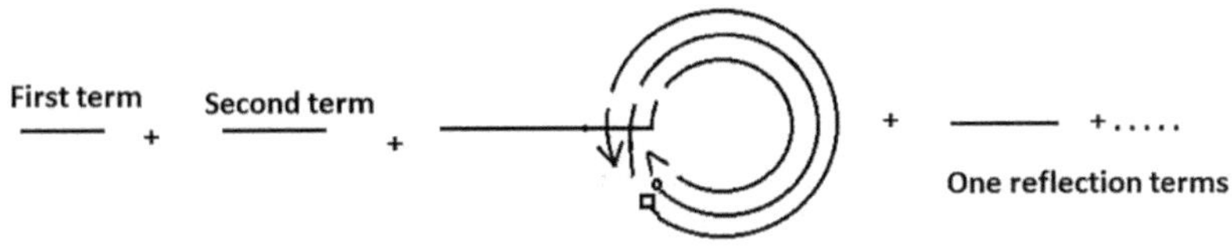

Fig. 1.14 It depicts the Feynman paths leading to the series in Eq. 1.85

One reflection terms (see Fig. 1.14)

$$\psi_1^{tt}(l^-) = te^{iql+i\alpha}r' + te^{iql+i\alpha}r'e^{iql-i\alpha}t' + ...$$

$$= \frac{tr'e^{iql+i\alpha}}{1 - t'e^{iql-i\alpha}} \tag{1.85}$$

Two reflection terms

$$\psi_2^{tt}(l^-) = te^{iql+i\alpha}r'e^{iql-i\alpha}re^{iql+i\alpha} + te^{iql+i\alpha}r'e^{iql-i\alpha}re^{iql+i\alpha}t'e^{iql+i\alpha} + ...$$

$$= \frac{tr're^{3iql+i\alpha}}{1 - t'e^{iql+i\alpha}} \tag{1.86}$$

Three reflection terms

$$\psi_3^{tt}(l^-) = te^{iql+i\alpha}r'e^{iql-i\alpha}re^{iql+i\alpha}r' + ...e^{iql-i\alpha}t' + ...$$

$$= \frac{trr'^2e^{3iql+i\alpha}}{1 - t'e^{iql-i\alpha}} \tag{1.87}$$

Four reflection terms

$$\psi_4^{tt}(l^-) = te^{iql+i\alpha}r'e^{iql-i\alpha}re^{iql+i\alpha}r'e^{iql-i\alpha}re^{iql+i\alpha} + ...t'e^{iql+i\alpha} + ...$$

$$= \frac{tr^2r'^2e^{5iql+i\alpha}}{1 - t'e^{iql+i\alpha}} \tag{1.88}$$

Five reflection terms

$$\psi_5^{tt}(l^-) = te^{iql+i\alpha}r'e^{iql-i\alpha}re^{iql+i\alpha}r'e^{iql-i\alpha}re^{iql+i\alpha}r' + ...e^{iql-i\alpha}t' + ...$$

$$= \frac{tr'^3r^2e^{5iql+i\alpha}}{1 - t'e^{iql-i\alpha}} \tag{1.89}$$

So the contribution to the wave-function inside the ring at the point $x = l^-$ coming from these diagrams will again form two geometric series to give

$$\psi^{tt}(l^-) = [\psi_0^{tt}(l^-) + \psi_2^{tt}(l^-) + \psi_4^{tt}(l^-) + ...] + [\psi_1^{tt}(l^-) + \psi_3^{tt}(l^-) + \psi_5^{tt}(l^-) + ...]$$

$$= \frac{te^{iql+i\alpha}}{1 - t'e^{iql+i\alpha}}(1 + r'r e^{2iql} + r^2 r'^2 e^{4iql} + ...)$$

$$+ \frac{tr'e^{iql+i\alpha}}{1 - t'e^{iql-i\alpha}}(1 + rr'e^{2iql} + r^2 r'^2 e^{4iql} + ...)$$

$$= \frac{te^{iql+i\alpha}}{1 - t'e^{iql+i\alpha}} \frac{1}{1 - rr'e^{2iql}} + \frac{tr'e^{iql+i\alpha}}{1 - t'e^{iql-i\alpha}} \frac{1}{1 - rr'e^{2iql}} \tag{1.90}$$

Thus the resultant wave-function at $x = l^-$ is (from Eqs. 1.83 and 1.90)

$$\psi(l^-) = \psi^{tb}(l^-) + \psi^{tt}(l^-)$$

$$= \frac{t'}{1 - t'e^{iql-i\alpha}} \frac{1}{1 - rr'e^{2iql}} + \frac{t're^{2iql}}{1 - t'e^{iql+i\alpha}} \frac{1}{1 - rr'e^{2iql}}$$

$$+ \frac{te^{iql+i\alpha}}{1 - t'e^{iql+i\alpha}} \frac{1}{1 - rr'e^{2iql}} + \frac{tr'e^{iql+i\alpha}}{1 - t'e^{iql-i\alpha}} \frac{1}{1 - rr'e^{2iql}}$$

$$= \frac{te^{iql+i\alpha}}{1 - rr'e^{2iql}}[\frac{1}{1 - t'e^{iql+i\alpha}} + \frac{r'}{1 - t'e^{iql-i\alpha}}]$$

$$+ \frac{t'}{1 - rr'e^{2iql}}[\frac{1}{1 - t'e^{iql-i\alpha}} + \frac{re^{2iql}}{1 - t'e^{iql+i\alpha}}]$$

Note that this is again of the form $ae^{iql+i\alpha} + be^{-iql}$ where

$$a = \frac{t}{1 - rr'e^{2iql}}\left[\frac{1}{1 - t'e^{iql+i\alpha}} + \frac{r'}{1 - t'e^{iql-i\alpha}}\right] \tag{1.91}$$

$$b = \frac{t'e^{iql}}{1 - rr'e^{2iql}}\left[\frac{1}{1 - t'e^{iql-i\alpha}} + \frac{re^{2iql}}{1 - t'e^{iql+i\alpha}}\right] \tag{1.92}$$

Note that once we substitute $r' = re^{2iql}$ (explained below) then the expressions for a and b become the same in absence of magnetic field. Also note that it will give the same current as that from Eqs. 1.74 and 1.75 at $x = 0^+$.

1.6 Scattering Phase Shift and DOS

Note that in Fig. 1.1 the coordinate in the right lead is taken as x' with the $x' = 0$ point depicted in the figure. $x = 0$ point is also depicted. One can write the wave-function on the right as

$$\psi(x \geq l) = te^{ik(x-l)} = te^{-ikl}e^{ikx} = t'e^{ikx} \tag{1.93}$$

In this picture in absence of potential $t' = 1$, as the wave-function at $x = l$ should be e^{ikl}. So the entity $t = e^{ikl}$ in absence of scatterer. So in the absence of the potential, the phase of t, i.e., θ_t will be

$$\theta_t = kl$$

Therefore, its derivative with respect to incident energy E will be

$$\frac{d\theta_t}{dE} = \frac{d\theta_t}{dk}\frac{dk}{dE} = \frac{d(kl)}{dk}\frac{1}{\frac{dE}{dk}} = l\frac{1}{\frac{d}{dk}\frac{\hbar^2k^2}{2m_0}} = \frac{2m_0 l}{\hbar^2}\frac{1}{2k} = \frac{m_0 l}{\hbar^2 k} = \frac{2\pi l}{h}\frac{1}{\frac{\hbar k}{m_0}} = 2\pi l\frac{1}{hv}$$

The factor 2 comes because when we write the dispersion relation as $E = \frac{\hbar^2k^2}{2m_0}$ then both $\pm k$ states are taken into account. Therefore,

$$\frac{d\theta_t}{dE} = \pi\rho_0 \tag{1.94}$$

where $\rho_0 = \frac{1}{hv}$ per unit length is 1D DOS in absence of scatterer. Up to this the potential U is zero. We can either call t' to be the scattering amplitude or call t to be the scattering amplitude. In the first picture the scattering phase shift is the phase of t' and it is zero while in the second picture the scattering phase shift is the phase of t which is $\theta_t = kl$. Difference between t' and t is just a shift of origin.

In presence of a potential $U \neq 0$ again depending on where we fix the origin, we may work with t or with t'. In presence of scatterer one can show (see Chap. 3)

$$\frac{d\theta_t}{dE} \approx \pi\rho \tag{1.95}$$

As is evident from Eq. 1.93 that

$$\theta_{t'} = \theta_t - kl \tag{1.96}$$

This implies from Eqs. 1.94 and 1.95

$$\frac{d\theta_{t'}}{dE} \approx \pi(\rho - \rho_0) \tag{1.97}$$

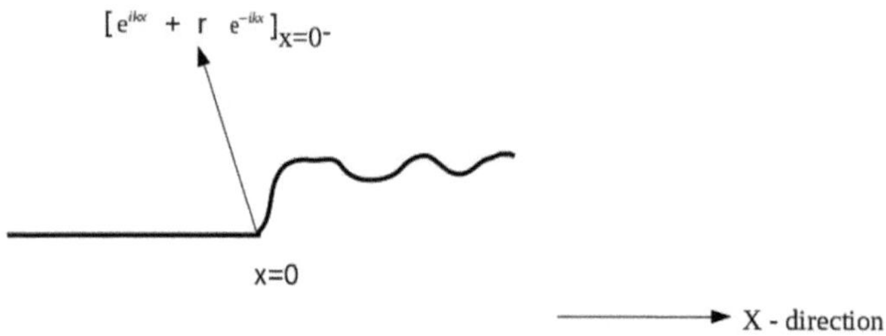

Fig. 1.15 The bold curve represents a 1D potential profile. X-direction is shown and choice of origin is also shown. This choice of origin implies the wave-function that is matched at the origin is pointed with an arrow

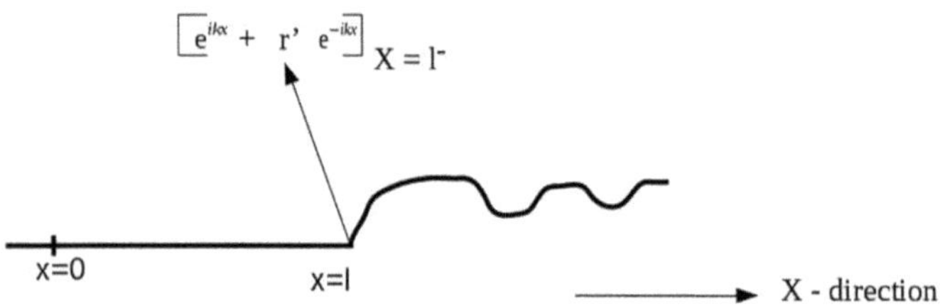

Fig. 1.16 The same scattering problem as that in Fig. 1.15 but the origin is shifted which changes the wave-function that determines the boundary conditions

We will work with $te^{ik(x-l)}$ or $te^{ikx'}$, meaning x' is a new coordinate system that is 0 at $x = l$. It is just a change of variable but it makes keeping track of phase changes with respect to the initial phase easier. In fact if the system has many such constituent parts joined by junctions, then it is suitable to assign a new origin for each region. However, the reflection and transmission at the different junctions in the Feynman path approach have to be appropriately corrected for the necessary phase factor required. For example, in Fig. 1.4, t is transmission amplitude of the three prong potential and r is the reflection amplitude. A simple calculation shows $t = 2/3$ and $r = -1/3$. This is the special case for the system solved in detail in Chap. 5. However, when the reflection occurs at $x = l$ then the reflection amplitude should be written as r' and it is related to r as follows:

$$\left\langle re^{-ik0^-} \middle| e^{ik0^-} \right\rangle = \left\langle r'e^{-ikl^-} \middle| e^{ikl^-} \right\rangle \tag{1.98}$$

This explains the substitution used in Eq. 1.76. r' is the reflection amplitude at $x = l$ while r is the reflection amplitude at $x = 0$. By symmetry, the amplitude of r and r' will be the same and equal to $-\frac{1}{3}$ but their phases will differ. This can be further understood by comparing Figs. 1.15 and 1.16. For the same reason Eqs. 1.91 and 1.92 are explained since only anticlockwise moving $-k$ states are reflected at $x = 0$.

1.7 A Simple Understanding of Mesoscopic Dissipation

Note that in Fig. 1.3 we have depicted the wave-function in different regions of the sample and we have not written the wave-function in the reservoir. Then we said that the electrons are injected by the reservoir into the system and wave-function for these injected electrons is e^{ikx} which is justified by Eq. 1.5. Alternately, we can write the incident electron wave-function as $(\frac{2\pi m_0}{\hbar^2 k})^{1/2}e^{ikx}$ and this has been explained with Eq. 1.10. For the latter the boundary conditions give us three equations that are written in Eqs. 1.44 and 1.45. The three equations can be represented in matrix form and by matrix inversion we can determine the three unknowns that are a, b, and R. Current inside the ring is given by Eq. 1.50 which is an equilibrium current known as persistent current and it is plotted in Fig. 1.17 for a ring of unit length at a fixed flux of $0.45\Phi_0$. We can see a series of narrow peaks in the current. The first peak carries a negative current or a diamagnetic current and the peaks alternate between diamagnetic and paramagnetic. These peaks correspond to resonant states and we have thus given a detailed account of how we get these resonances in textbooks and how we get them here by not writing the wave-function in the reservoir and imposing some axioms (or physical conditions) on the incident wave. Note that we have a different wave vector q inside the ring (see Eq. 1.38) and by taking a stronger attractive potential U inside the ring we can make these resonances narrower. The width of these resonances is a depiction of the fact that the states in the ring can leak out into the reservoir and this is the cause of dissipation in mesoscopic physics. In the Landauer two probe conductance set up we have two reservoirs that create these resonances and hence dissipation or loss. In Fig. 5.2d again one can see similar resonances in scattering cross section that is well discussed in quantum mechanics text books. Mesoscopic currents (see Eq. 1.11) and conductance (see Eq. 2.24) are related to this cross section. It is generally believed that as the width of these resonances tend to zero, we will get the eigen-energies of the system for which there is no decay and no loss. The eigen-energy problem for a closed ring will be solved in Chap. 2. In the von-Neumann approach, the eigen-energies can be obtained from the singularities of R as will be demonstrated in Chap. 4 in a more general way (see Eqs. 4.80 and 4.94). Thus in the von-Neumann approach the scattering problem or the open system is more general than the closed system. In this book we will show that one may interpret states of the system are created with the winding number of the phase of R. This will establish known facts like density of states, etc., and also explain several ill understood concepts in mesoscopic physics that will prove the reality of time travel.

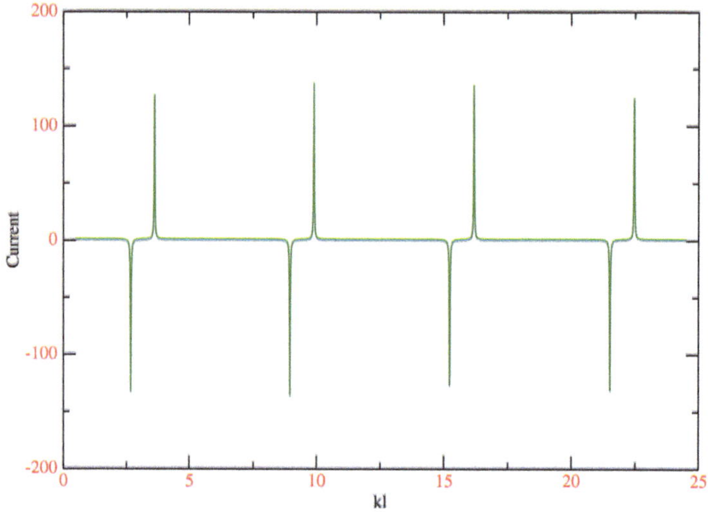

Fig. 1.17 Persistent current in the system of Fig. 1.3 being carried by a series of resonant states

1.8 Appendix: Hamiltonian of an Electron in Magnetic Field

The Euler-Lagrange (EL) equation of motion is

$$\frac{d}{dt}\left(\frac{\partial L}{\partial v_x}\right) - \frac{\partial L}{\partial x} = 0 \tag{1.99}$$

where v_x, x are the generalized velocities and coordinates that are independent of each other. It is an alternate way of arriving at the equation of motion of a particle and has certain advantages over Newton's equation of motion. First is that EL Equation depends on generalized coordinates and so independent of the choice of the coordinate system. For example, in case of Newton's equation of motion, if we want to change from Cartesian coordinates to polar coordinates, the algebra is quite cumbersome. Second is that the EL Equation is more suited to treat constraint forces. Thirdly, the EL Equation is more general and applicable to a much wider range of systems like LCR circuits, electromagnetic fields, etc., once we identify the generalized coordinates. Fourth is that in some cases, but not always, it gives us a way to obtain the quantum mechanical Hamiltonian from the classical Hamiltonian as will be the case here. We first show below that for an electron in a magnetic field the Lagrangian is

$$L = \frac{1}{2} m_0 v^2 - e_0 U + e_0 \vec{v} \cdot \vec{A}$$

Our starting point is that the Lorentz force acting on a charge e_0 moving with velocity $\vec{v}$ in an electric field $\vec{E}$ and magnetic field $\vec{B}$ is given by

$$\vec{F} = e_0\left[\vec{E} + \left(\vec{v} \times \vec{B}\right)\right] \tag{1.100}$$

From Maxwell's equations we know

$$\vec{E} = -\vec{\nabla}U - \frac{\partial \vec{A}}{\partial t} \text{ and } \vec{B} = \vec{\nabla} \times \vec{A} \tag{1.101}$$

where $\vec{A}$ is magnetic vector potential, $\vec{B}$ is magnetic field, and U is electric scalar potential. Substituting from Eq. 1.101 in Eq. 1.100

$$\vec{F} = e_0\left[-\vec{\nabla}U - \frac{\partial \vec{A}}{\partial t} + \vec{v} \times \vec{\nabla} \times \vec{A}\right] \tag{1.102}$$

Let us consider only the x-component of the force $\vec{F}$.

$$F_x = e_0\left[-\frac{\partial U}{\partial x} - \frac{\partial A_x}{\partial t} + \left(\vec{v} \times \vec{\nabla} \times \vec{A}\right)_x\right] \tag{1.103}$$

Now,

$$\vec{\nabla} \times \vec{A} = \begin{bmatrix} \hat{i} & \hat{j} & \hat{k} \\ \frac{\partial}{\partial x} & \frac{\partial}{\partial y} & \frac{\partial}{\partial z} \\ A_x & A_y & A_z \end{bmatrix} \tag{1.104}$$

$$= \hat{i}\left(\frac{\partial A_z}{\partial y} - \frac{\partial A_y}{\partial z}\right) + \hat{j}\left(\frac{\partial A_x}{\partial z} - \frac{\partial A_z}{\partial x}\right) + \hat{k}\left(\frac{\partial A_y}{\partial x} - \frac{\partial A_x}{\partial y}\right) \tag{1.105}$$

So

$$\vec{v} \times \left(\vec{\nabla} \times \vec{A}\right) = \begin{bmatrix} \hat{i} & \hat{j} & \hat{k} \\ v_x & v_y & v_z \\ \frac{\partial A_z}{\partial y} - \frac{\partial A_y}{\partial z} & \frac{\partial A_x}{\partial z} - \frac{\partial A_z}{\partial x} & \frac{\partial A_y}{\partial x} - \frac{\partial A_x}{\partial y} \end{bmatrix} \tag{1.106}$$

Therefore

$$\left(\vec{v} \times \vec{\nabla} \times \vec{A}\right)_x = v_y\frac{\partial A_y}{\partial x} + v_z\frac{\partial A_z}{\partial x} - v_y\frac{\partial A_x}{\partial y} - v_z\frac{\partial A_x}{\partial z} \tag{1.107}$$

Adding and subtracting $v_x\frac{\partial A_x}{\partial x}$ on RHS and considering that v_x and x are generalized coordinates independent of each other

$$\left(\vec{v}\times\vec{\nabla}\times\vec{A}\right)_x = v_x\frac{\partial A_x}{\partial x}+v_y\frac{\partial A_y}{\partial x}+v_z\frac{\partial A_z}{\partial x}-v_x\frac{\partial A_x}{\partial x}-v_y\frac{\partial A_x}{\partial y}-v_z\frac{\partial A_x}{\partial z} \tag{1.108}$$

$$=\left(v_x\frac{\partial A_x}{\partial x}+v_y\frac{\partial A_y}{\partial x}+v_z\frac{\partial A_z}{\partial x}\right)-\left(v_x\frac{\partial A_x}{\partial x}+v_y\frac{\partial A_x}{\partial y}+v_z\frac{\partial A_x}{\partial z}\right) \tag{1.109}$$

$$=\left(\frac{\partial(v_xA_x)}{\partial x}+\frac{\partial(v_yA_y)}{\partial x}+\frac{\partial(v_zA_z)}{\partial x}\right)-\left(v_x\frac{\partial A_x}{\partial x}+v_y\frac{\partial A_x}{\partial y}+v_z\frac{\partial A_x}{\partial z}\right) \tag{1.110}$$

$$=\left(\frac{\partial}{\partial x}(\vec{v}\cdot\vec{A})\right)-\left(v_x\frac{\partial A_x}{\partial x}+v_y\frac{\partial A_x}{\partial y}+v_z\frac{\partial A_x}{\partial z}\right) \tag{1.111}$$

$$=\left(\frac{\partial}{\partial x}(\vec{v}\cdot\vec{A})\right)-\left(\frac{dx}{dt}\frac{\partial A_x}{\partial x}+\frac{dy}{dt}\frac{\partial A_x}{\partial y}+\frac{dz}{dt}\frac{\partial A_x}{\partial z}+\frac{dt}{dt}\frac{\partial A_x}{\partial t}-\frac{dt}{dt}\frac{\partial A_x}{\partial t}\right) \tag{1.112}$$

$$=\left(\frac{\partial}{\partial x}(\vec{v}\cdot\vec{A})\right)-\left(\frac{dA_x}{dt}-\frac{dt}{dt}\frac{\partial A_x}{\partial t}\right) \tag{1.113}$$

Therefore

$$\left(\vec{v}\times\vec{\nabla}\times\vec{A}\right)_x=\frac{\partial}{\partial x}\left(\vec{v}\cdot\vec{A}\right)-\left(\frac{dA_x}{dt}-\frac{\partial A_x}{\partial t}\right) \tag{1.114}$$

Substituting this in 1.103 we have

$$F_x=e_0\left[-\frac{\partial U}{\partial x}-\frac{\partial A_x}{\partial t}+\frac{\partial}{\partial x}\left(\vec{v}\cdot\vec{A}\right)-\frac{dA_x}{dt}+\frac{\partial A_x}{\partial t}\right] \tag{1.115}$$

$$\text{or}\quad F_x=e_0\left[-\frac{\partial}{\partial x}\left(U-\vec{v}\cdot\vec{A}\right)-\frac{d}{dt}A_x\right] \tag{1.116}$$

$$\text{or}\quad F_x=e_0\left[-\frac{\partial}{\partial x}\left(U-\vec{v}\cdot\vec{A}\right)-\frac{d}{dt}\left(\frac{\partial}{\partial v_x}(v_xA_x+v_yA_y+v_zA_z)\right)\right] \tag{1.117}$$

$$\text{or}\quad F_x=e_0\left[-\frac{\partial}{\partial x}\left(U-\vec{v}\cdot\vec{A}\right)-\frac{d}{dt}\left(\frac{\partial}{\partial v_x}(\vec{v}\cdot\vec{A})\right)\right] \tag{1.118}$$

$$\text{or}\quad F_x=e_0\left[-\frac{\partial}{\partial x}\left(U-\vec{v}\cdot\vec{A}\right)+\frac{d}{dt}\left(\frac{\partial}{\partial v_x}(U-\vec{v}\cdot\vec{A})\right)\right] \tag{1.119}$$

Scalar U is not a function of v_x. Thus with

$$e_0\left(U-\vec{v}\cdot\vec{A}\right)\rightarrow \mathrm{U} \tag{1.120}$$

as the generalized potential

$$F_x = \left[\frac{d}{dt}\frac{\partial}{\partial v_x}(U) - \frac{\partial}{\partial x}(U)\right] \tag{1.121}$$

$$\text{or } F_x = \frac{d}{dt}\frac{\partial U}{\partial v_x} - \frac{\partial U}{\partial x} \tag{1.122}$$

For $T = \frac{1}{2}m_0 v_x^2$, Newton's equation can be written as

$$\frac{d}{dt}\left(\frac{\partial T}{\partial v_x}\right) - \frac{\partial T}{\partial x} = F_x \tag{1.123}$$

From Eqs. 1.122 and 1.123 we get

$$\frac{d}{dt}\frac{\partial}{\partial v_x}(T - U) - \frac{\partial}{\partial x}(T - U) = 0 \tag{1.124}$$

Thus we get the Lagrangian as (see Eq. 1.120)

$$L = T - U = \frac{1}{2}m_0 v_x^2 - e_0 U + e_0 v_x A_x \tag{1.125}$$

To get the Hamiltonian from the Lagrangian

$$p_x = \frac{\partial L}{\partial v_x} = m_0 v_x + e_0 A_x \Rightarrow v_x = \frac{p_x - e_0 A_x}{m_0} \tag{1.126}$$

Hence we write

$$\vec{v} = \frac{\vec{p} - e_0\vec{A}}{m_0} \tag{1.127}$$

It is to be noted that in presence of magnetic field, p_x (and not $m_0 v_x$) is conjugate variable for x. The kinetic momentum is still $m_0 v_x$ and $e_0 A_x$ is the field momentum. The Hamiltonian is obtained by Legendre transformation as

$$\begin{aligned}
& H = \sum p_i \dot{q}_i - L \\
\text{or } & H = \vec{p}\cdot\vec{v} - L \\
\text{or } & H = \frac{\vec{p}\cdot\left(\vec{p} - e_0\vec{A}\right)}{m_0} - \frac{1}{2}m_0\frac{\left(\vec{p} - e_0\vec{A}\right)^2}{m_0^2} + e_0 U - e_0\vec{A}\cdot\left(\frac{\vec{p} - e_0\vec{A}}{m_0}\right) \\
\text{or } & H = \frac{p^2 - \vec{p}\cdot e_0\vec{A} - e_0\vec{A}\cdot\vec{p} + e_0^2 A^2}{m_0} - \frac{1}{2}m_0\frac{\left(\vec{p} - e_0\vec{A}\right)^2}{m_0^2} + e_0 U \\
\text{or } & H = \frac{1}{2m_0}\left(\vec{p} - e_0\vec{A}\right)^2 + e_0 U
\end{aligned}$$

From this we can understand Eq. 1.13.

References

1. Datta S (1995) Electronic transport in mesoscopic systems. Cambridge University Press, reprinted in 1999
2. Pastawski HM (1992) Phys. Rev. B 46:4053
3. Bonneau G et al (2001) Am. J. Phys 69:322
4. Mukherjee S et al (2011) Physica E 44:62
5. Buttiker M (1985) Phys. Rev. B 32:1846(R)
6. Bluhm H et al (2009) Phys. Rev. Lett. 102(13):136802 and references therein
7. Mukherjee S et al (2020) Physica E 118:113933

Chapter 2
Closed Systems

Abstract As it is known that quantum mechanical states exhibit an even-odd effect called parity effect, in this chapter we will discuss some examples wherein it has direct consequences on experimental observations. Magnetization of mesoscopic rings introduced in previous chapter is an ideal avenue to explore parity effect because the magnetization of a closed ring can be measured non-invasively without attaching leads. Then we will show that Fano resonances can non-trivially change this parity effect by introducing a new phase to the electron wave-function. Finally, in the end we will acquaint the reader with what is a general mesoscopic system in a schematic manner to help understand the connection between this new phase and mesoscopic properties in general. Finally, we will provide a flow chart of logic that will make a strong connection between von-Neumann quantum mechanics and physics of mesoscopic systems. This is important because that is known to be the correct version of quantum mechanics and in Chaps. 5 and 6 we will prove that time travel is a reality in the mesoscopic regime and can be achieved with present day technology. The flow chart will also make it clear what we mean by theoretical experiments that will be presented in Chap. 7 and why they are more authentic than practical ones.

In the previous chapter we introduced the reader to open systems with a simple example that of a ring connected to an electron reservoir. In this chapter we will present the closed version of the same system. While mesoscopic open systems are understood by setting up a scattering problem, closed Hamiltonian systems are understood in terms of eigen-states and eigen-energies. One practical way to see the connection between eigen-states and scattering states is via Friedel sum rule (FSR) that was preliminarily introduced in Chap. 1, Eq. 1.95. FSR like formulas are extensively used in condensed matter systems that are usually semi-classical. Hence we see an approximate equality in Eq. 1.95 which will be properly clarified in Chap. 3. The simple principle behind arriving at this equation is that by estimating scattering phase change in a range (say a range of incident energy), we can figure out the number of allowed eigen-states of the system, in that range. Every π phase change corresponds to a new eigen-function of the potential. The phase change introduces new nodes in the eigen-function and as the nodes increase, the potential

P. S. Deo and K. Meena, *Time Travel: A Reality in Mesoscopic Physics*,
https://doi.org/10.1007/978-981-96-1136-2_2

can accommodate more and more eigen-states. As a result, one finds that number of states below a particular state can be found by counting the number of nodes in the wave-function of that state and subtracting one, as the ground state has zero nodes. These facts will be explained in detail as we proceed. Thus mesoscopic systems give us a route to make a one is to one connection between loss-less eigen-states and states in an open system that exhibit dissipation and loss. Once dissipation and decoherence dominates, the system becomes classical and new laws and principles become effective. We usually learn to treat classical systems and quantum systems through completely different principles. The route to connect them can come from mesoscopic systems.

We will continue with the simple 1D ring for basic understanding and try to generalize in later chapters. Subsequently, a simple modification of the closed ring system will reveal some interesting effects that has only recently come to light and has a bearing on FSR. In case of small quantum systems, FSR type of formulas will break down for two reasons. (1) Quantum corrections: they will be significant as these small quantum systems cannot be treated semi-classically. However, people have thought about it before us to some extent. (2) Number of states is one more than the number of nodes is not always true: this not only changes our view of the first point above but leads to many dramatic consequences that require us to re-think the whole of physics in some sense. In Chap. 6 we will see that one can realize that signals can be sent in negative time (classically this means faster than the speed of light) within the single particle coherence length.

2.1 Parity Effect in Closed Ring

Consider an infinite square well potential of length L shown in Fig. 2.1. The quantum mechanical wave-functions are drawn by dashed lines and they exhibit a parity effect wherein states with an even number of nodes and that with an odd number of nodes alternate. This gives an explicit example of the statement that the number labeling a state or the index of a state is one more than the number of nodes. Now we can also draw an infinite square well potential like a one-dimensional line with perfectly reflecting dead ends. So the one-dimensional line can also be curled up to form a ring as shown in Fig. 2.2. An Aharonov-Bohm flux Φ can be placed in the center of the ring such that the electrons in the ring do not feel any magnetic field. The alternating even and odd state of the 1D line or the parity effect manifests in the states of the ring in a particular way as we will discuss now.

Consider a one-dimensional ring of length L. The quantum mechanical wave-function in the ring is that of a free electron $\psi = e^{ikx}$ where ψ is the solution to the 1D Schrodinger Eq. $-\frac{\hbar^2}{2m_0}\frac{d^2\psi}{dx^2} = E\psi$. The eigen-energies can be found from the single valuedness of the wave function that yield the condition

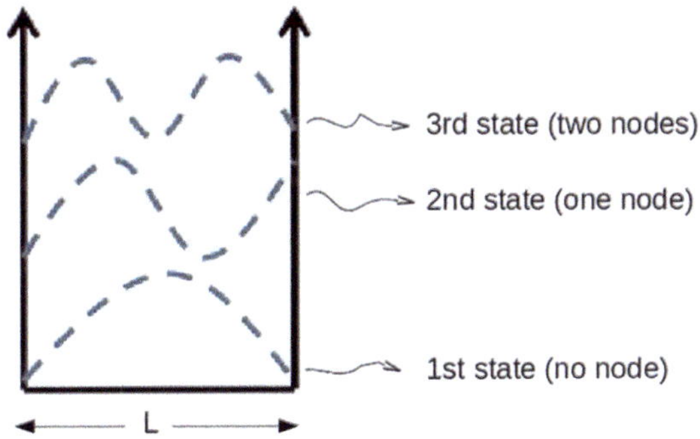

Fig. 2.1 An infinite potential well with states exhibiting parity effect

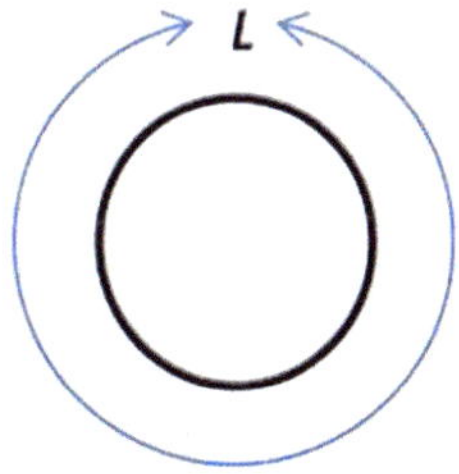

Fig. 2.2 A 1D ring of length L

$$\begin{aligned} e^{ik(x+L)} &= e^{ikx} \\ \text{or,}\ e^{ikL} &= 1 = e^{\pm i2n\pi} \qquad n = 0, 1, 2... \\ \text{or,}\ k &= \pm\frac{2n\pi}{L} \end{aligned} \tag{2.1}$$

$$E_n = \frac{\hbar^2 k_n^2}{2m_0} = \frac{2\hbar^2 n^2 \pi^2}{m_0 L^2} \tag{2.2}$$

Now if an Aharonov-Bohm flux is introduced piercing the ring (note that there is no magnetic field $\mathbf{B} = \nabla \times \mathbf{A}$ on the electrons in the ring as we keep the flux confined to a small area inside the ring) then the above condition changes to

$$\begin{aligned} e^{i(kL-\alpha)} &= 1 = e^{\pm i2n\pi} \qquad \text{where}\ \alpha = \frac{2\pi\Phi}{\Phi_0} \\ \text{or,}\ k &= \frac{\pm 2n\pi + \alpha}{L} \end{aligned} \tag{2.3}$$

where $\alpha = \frac{2\pi\Phi}{\Phi_0}$ can be positive as well as negative depending on the direction of the field, where $\Phi_0 = \frac{hc}{e_0}$.

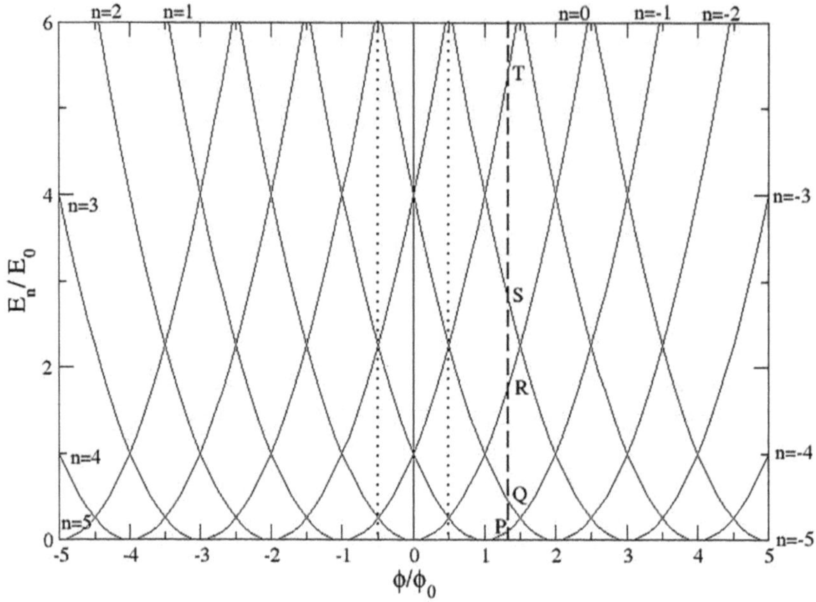

Fig. 2.3 The dotted lines enclose the first Brillouin zone. The dashed line is corresponding to $\frac{\Phi}{\Phi_0} = 1.34$

$$\begin{aligned} E_n &= \frac{\hbar^2 k_n^2}{2m_0} = \frac{\hbar^2}{2m_0} \frac{(\pm 2n\pi + \alpha)^2}{L^2} \\ &= \frac{\hbar^2}{2m_0 L^2} (\pm 2n\pi + \alpha)^2 \\ &= \frac{h^2}{2m_0 L^2} \left(n + \frac{\Phi}{\Phi_0} \right)^2 \end{aligned} \tag{2.4}$$

where $n = 0, \pm 1, \pm 2, \ldots..$

The single particle levels E_n are drawn in Fig. 2.3. As is evident from Eq. 2.4 the single particle eigen-energy E_n is parabolic in $\frac{\Phi}{\Phi_0}$. E_0 corresponding to $n = 0$ has minimum at $\frac{\Phi}{\Phi_0} = 0$ (see Eq. 2.4). E_1 corresponding to $n = +1$ is identical to E_0 but minimum displaced to $\frac{\Phi}{\Phi_0} = -1$. E_{-1} corresponding to $n = -1$ is identical to E_0 but minimum displaced to $\frac{\Phi}{\Phi_0} = +1$ and so on. The region between the two dotted lines, i.e., $|\frac{\Phi}{\Phi_0}| \leq \frac{1}{2}$ defines the Brillouin zone as by repeating it horizontally we get rest of the spectrum. Now if we fix the flux $\frac{\Phi}{\Phi_0} = 1.34$ then the eigen energies of the ring are at the energies corresponding to the points P, Q, R, S, T, etc. The state at P is a diamagnetic state as its energy increases with flux and that at Q is a paramagnetic state as its energy decreases with flux. The slope of the single particle energy levels implies that they carry a current. The current carried by a particular level n is $I_n = -c\frac{\partial E_n}{\partial \Phi}$. This is a purely equilibrium current called persistent current and it occurs because the flux lifts the degeneracy between the clockwise moving and anticlockwise moving states. One can see that at specific values of $\Phi = \frac{n}{2}\Phi_0$ these two time reversed states become degenerate and their currents cancel each other. Thus the paramagnetic and diamagnetic states alter just as in Fig. 2.1 the even and odd states alter. This is just the parity effect of the 1D line manifesting in a different way. States with an even number of nodes in the 1D line become diamagnetic in a 1D ring, while states with an odd number of nodes in the 1D line become paramagnetic in the 1D ring. Counting of nodes can lead to very general arguments to state that even in presence of many body effects and disorder, alternate many body states will carry opposite currents. These arguments are due to Leggett and outlined below.

Variational argument for parity effect in many electron states: The generalized electronic Hamiltonian for N electrons in a one-dimensional ring, described by angular coordinate θ is

$$H\left(\theta_1, \cdots \theta_N\right) = \sum_{i=1}^{N} V(\theta_i) + \sum_{i \neq j}^{N} U(\theta_i - \theta_j) + \sum_{i=1}^{N} \frac{\left[p_i - \frac{eA(\theta_i)}{c}\right]^2}{2m_0} \tag{2.5}$$

where $V(\theta_i)$ is a one body potential and $U(\theta_i - \theta_j)$ is a two body potential. Denoting the many body wave-function as $\Psi(\theta_1, ..., \theta_N)$, energy E is

$$E = \left\langle \Psi\left(\theta_1, \cdots \theta_N\right) | H | \Psi\left(\theta_1, \cdots \theta_N\right) \right\rangle \tag{2.6}$$

$$= \int \sum_i V(\theta_i)\Psi^*\left(\theta_1, \cdots \theta_N\right) \Psi\left(\theta_1, \cdots \theta_N\right) d\theta_1 \cdots d\theta_N$$

$$+ \int \sum_{i \neq j} U(\theta_i - \theta_j)\Psi^*\left(\theta_1, \cdots \theta_N\right) \Psi\left(\theta_1, \cdots \theta_N\right) d\theta_1 \cdots d\theta_N + KE$$

$$= PE + KE \tag{2.7}$$

Here, KE refers to kinetic energy and PE consisting of the first two terms refer to potential energy. Now we want to construct a variational wave-function. Let the variational wave-function be

$$\Psi_{var}\left(\varphi, \theta_1, \cdots \theta_N\right)$$

φ is a variational parameter such that by tuning this parameter this wave-function can be made an appropriate many body Fermionic wave-function for the system in consideration. Without any loss of generality, we can say that by fixing the reference point, for $\varphi = 0$, Ψ_{var} becomes the Fermionic wave-function. We must ensure that Ψ_{var} respects the symmetries of the original wave-function. So, Ψ_{var} should satisfy the following conditions.

1. If any one of $\theta_1, \theta_2, \cdots \theta_N$ changes by 2π, Ψ_{var} should pick up a phase χ ($\chi = \frac{2\pi\Phi}{\Phi_0}$). Here Φ is the total flux enclosed by the ring and $\Phi_0 = \frac{hc}{e_0}$. Note that by this choice of Ψ_{var} we are working with the gauge transformed Hamiltonian where KE operator is $\frac{p_i^2}{2m_0}$ and not $\frac{(p_i - eA_i)^2}{2m_0}$. So,

$$\Psi_{var}\left(\varphi, \theta_1, \cdots \theta_N\right) = e^{i\chi(\theta_1, \cdots \theta_N)} \Psi\left(\varphi, \theta_1, \cdots \theta_N\right) \tag{2.8}$$

where

$$-\chi\left(\theta_1, \cdots, \theta_i = 0, \cdots, \theta_N\right) + \chi\left(\theta_1, \cdots, \theta_i = 2\pi, \cdots, \theta_N\right) = \frac{2\pi\Phi}{\Phi_0}$$

2. Ψ_{var} should be anti-symmetric for $\varphi = 0$, i.e.,

$$\Psi_{var}\left(\varphi = 0, \theta_1, \theta_2, \cdots \theta_N\right) = -\Psi_{var}\left(\varphi = 0, \theta_2, \theta_1, \cdots \theta_N\right) \tag{2.9}$$

or for any other interchange of θ_i and θ_j. Therefore,

$$\begin{aligned} & e^{i\chi(\theta_1, \theta_2, \cdots \theta_N)} \Psi\left(0, \theta_1, \theta_2, \cdots \theta_N\right) \\ &= -e^{i\chi(\theta_2, \theta_1, \cdots \theta_N)} \Psi\left(0, \theta_2, \theta_1, \cdots \theta_N\right) \\ &= -e^{i\chi(\theta_2, \theta_1 \cdots \theta_N)} \left[-\Psi\left(0, \theta_1, \theta_2 \cdots \theta_N\right)\right] \qquad \text{(from Eqs. 2.8 and 2.9)} \end{aligned}$$

as $\Psi\left(0, \theta_2, \theta_1 \cdots \theta_N\right)$ is a Fermionic wave-function at $\varphi = 0$. So

$$\chi\left(\theta_1, \theta_2, \cdots \theta_N\right) = \chi\left(\theta_2, \theta_1 \cdots \theta_N\right) \tag{2.10}$$

or for any other interchange of θ_i and θ_j. That is χ is a symmetric function of its variables.

As the initial phase is arbitrary, we choose $\Psi\left(0, \theta_1, \theta_2 \cdots \theta_N\right)$ to be real. For non-zero flux as well as for non-zero φ, Ψ_{var} is complex.

However, $\Psi_{var}(\Phi = 0)$ and $\Psi_{var}(\Phi \neq 0)$ have the same potential energy, as the potential energy depends on $|\Psi_{var}|^2$. Kinetic energy with these gauge transformed wave-functions is

$$KE = \text{Re}\langle\Psi_{var}(\Phi)| \sum_i^N \frac{p_i^2}{2m_0} |\Psi_{var}(\Phi)\rangle \tag{2.11}$$

If we use the exact wave-function instead of a Ψ_{var} in Eq. 2.11 then K.E. can only be a real number. The purpose of Ψ_{var} is to give a number as close to that as possible, which means it can have an imaginary part as well but the real part will be close to or exact to the actual K.E. Thus

$$KE = \text{Re} \int \Psi^*_{var} \sum_i \left(-\frac{\hbar^2}{2m_0} \right) \frac{\partial^2}{\partial \theta_i^2} \Psi_{var} d\theta_1 \cdots d\theta_N \tag{2.12}$$

Since $\Psi_{var}(\Phi \neq 0) = \Psi_{var}(\Phi = 0)e^{i\chi} = \Psi_{or}e^{i\chi}$ (say), implying $\Psi_{or} = \Psi_{var}(\varphi = 0,$ $\Phi = 0)$ is real.

$$\begin{aligned} KE &= -\frac{\hbar^2}{2m_0} \sum_i \text{Re} \int \Psi^*_{var} \frac{\partial}{\partial \theta_i} \frac{\partial}{\partial \theta_i} \Psi_{or} e^{i\chi} d\theta_1 \cdots d\theta_N \\ &= -\frac{\hbar^2}{2m_0} \sum_i \text{Re} \int \Psi^*_{var} \frac{\partial}{\partial \theta_i} \left[\Psi_{or} i \frac{\partial \chi}{\partial \theta_i} e^{i\chi} + e^{i\chi} \frac{\partial \Psi_{or}}{\partial \theta_i} \right] d\theta_1 \cdots d\theta_N \\ &= -\frac{\hbar^2}{2m_0} \sum_i \text{Re} \int \Psi^*_{var} \left[\Psi_{or} i \frac{\partial \chi}{\partial \theta_i} i \frac{\partial \chi}{\partial \theta_i} e^{i\chi} + \Psi_{or} i e^{i\chi} \frac{\partial^2 \chi}{\partial \theta_i^2} + \frac{\partial \Psi_{or}}{\partial \theta_i} i \frac{\partial \chi}{\partial \theta_i} e^{i\chi} \right. \\ &\quad \left. + e^{i\chi} \frac{\partial^2 \Psi_{or}}{\partial \theta_i^2} + e^{i\chi} i \frac{\partial \chi}{\partial \theta_i} \frac{\partial \Psi_{or}}{\partial \theta_i} \right] d\theta_1 \cdots d\theta_N \\ &= -\frac{\hbar^2}{2m_0} \sum_i \text{Re} \int \left[-\Psi^*_{or} \Psi_{or} \left(\frac{\partial \chi}{\partial \theta_i} \right)^2 + \Psi^*_{or} \Psi_{or} i \frac{\partial^2 \chi}{\partial \theta_i^2} + \Psi^*_{or} i \frac{\partial \chi}{\partial \theta_i} \frac{\partial \Psi_{or}}{\partial \theta_i} \right. \\ &\quad \left. + \Psi^*_{or} \frac{\partial^2 \Psi_{or}}{\partial \theta_i^2} + i \frac{\partial \chi}{\partial \theta_i} \Psi^*_{or} \frac{\partial \Psi_{or}}{\partial \theta_i} \right] d\theta_1 \cdots d\theta_N \\ &= -\frac{\hbar^2}{2m_0} \sum_i \int \left[-|\Psi_{or}|^2 \left(\frac{\partial \chi}{\partial \theta_i} \right)^2 + \Psi^*_{or} \frac{\partial^2}{\partial \theta_i^2} \Psi_{or} \right] d\theta_1 \cdots d\theta_N \end{aligned} \tag{2.13}$$

Therefore change in kinetic energy due to Φ, in comparison with Eq. 2.12, is

$$\begin{aligned} &\frac{\hbar^2}{2m_0} \sum_i \int \left[|\Psi_{or}|^2 \left(\frac{\partial \chi(\theta_1, \theta_2, \cdots \theta_N)}{\partial \theta_i} \right)^2 \right] d\theta_1 \cdots d\theta_N \\ &\qquad = \frac{\hbar^2}{2m_0} N \int \left[|\Psi_{or}|^2 \left(\frac{\partial \chi}{\partial \theta_1} \right)^2 \right] d\theta_1 \cdots d\theta_N \geq 0 \end{aligned} \tag{2.14}$$

as the integrand is whole square. To explain the last step, we give an example with two terms in the sum.

$$\int |\Psi_{or}|^2 \left(\frac{\partial \chi(\theta_1, \theta_2, \cdots \theta_N)}{\partial \theta_1} \right)^2 d\theta_1 \cdots d\theta_N + \int |\Psi_{or}|^2 \left(\frac{\partial \chi(\theta_1, \theta_2, \cdots \theta_N)}{\partial \theta_2} \right)^2 d\theta_1 \cdots d\theta_N = J(\text{say})$$

In the second integration we put $\theta_1 \longrightarrow \theta_2$ and $\theta_2 \longrightarrow \theta_1$, and we know that

$$\chi(\theta_1, \theta_2, \cdots \theta_N) = \chi(\theta_2, \theta_1 \cdots \theta_N)$$

Therefore

$$J = \int |\Psi_{or}|^2 \left(\frac{\partial \chi(\theta_1, \theta_2, \cdots \theta_N)}{\partial \theta_1}\right)^2 d\theta_1 \cdots d\theta_N + \int |\Psi_{or}|^2 \left(\frac{\partial \chi(\theta_1, \theta_2, \cdots \theta_N)}{\partial \theta_1}\right)^2 d\theta_1 \cdots d\theta_N$$
$$= 2\int |\Psi_{or}|^2 \left(\frac{\partial \chi(\theta_1, \theta_2, \cdots \theta_N)}{\partial \theta_1}\right)^2 d\theta_1 \cdots d\theta_N$$

This explains why the ground state, second excited state, or any even state energy always increases with flux (for $\Phi < \Phi_0/2$), that is, it is a diamagnetic state. The first excited state however does not or to be precise does not necessarily do so. This follows from the following topological arguments. Suppose we have a ring with an even number of electrons, that is N is even. Now let us fix the coordinates of $N-1$ electrons which is an odd number of electrons and take the Nth electron, which we call the test electron, round the ring once. It has to pick up a phase of 2π when it comes back to its initial position because of the single valuedness of the many body wave-function. Note that this is a mathematical operation done at the level of representations that does not care whether one can fix the position of other electrons and whether the test electron can be physically taken round to its exact initial position. Every time it crosses an electron it picks up a phase of π due to anti-symmetric property of the many body wave-function. So when it goes round the ring once, it will pick up a π phase $(N-1)$ times or an odd number of times. In case N is odd with $(N-1)$ being an even number the π phase shift will occur an even number of times. This extra phase difference of π for the many body wave-function of even number of electrons with respect to that for odd number of electrons does not necessarily change continuously with any parameter like χ. So the first excited state, third excited state, or any odd state needs a non-symmetry-dictated-node (NSDN) to match its wave-function in going round the ring once and this makes them paramagnetic. To be precise any arbitrary function on either sides of a NSDN is automatically matched by the NSDN and so the odd electron states are at least non-diamagnetic.

2.2 Deviations in Parity Effect

Now let us consider two infinite square well potentials shown in Fig. 2.4 placed perpendicular to each other. We may also consider them as topological spaces, in which case the exact angle of meet between the two lines or intervals does not matter because topological spaces do not necessarily have a notion of angle. It is to be noted that these kinds of topological spaces are well studied in algebraic topology and differential geometry as graphs and algebraic varieties. Quantum phenomenon in such spaces is a motivation induced by mesoscopic physics and the ability to make such samples in the laboratory. We will also see that the quantum mechanical equations in 1D, leading to boundary condition equations, do not depend on the angle. However, in later chapters when we show physical realizations of this system curved

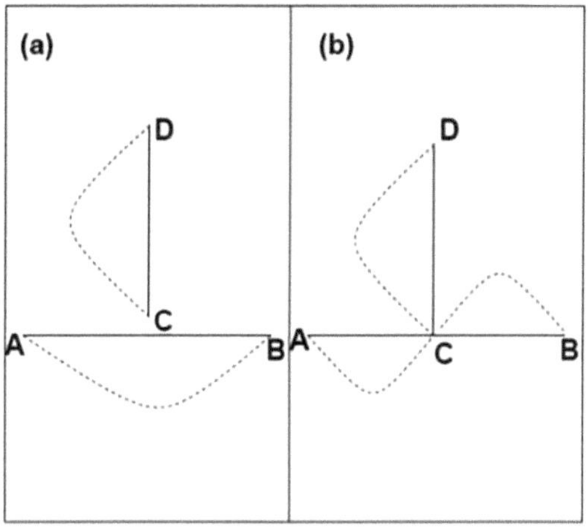

Fig. 2.4 Two infinite potential wells are placed perpendicular to each other. In **a** they are disjoint and in **b** they are joined

out of real 3D quantum wires then the two lines better be perpendicular for theoretical convenience and such systems have been fabricated in the laboratory confirming the results to be discussed below [K. Kobayashi, et al. [1], 2004]. One of the square wells is shown by the solid line AB and the other by the solid line CD. Initially they are decoupled from each other as shown in Fig. 2.4a. The ground state wave-functions in AB and CD are shown by the dotted lines and obviously both are even states. Subsequently, in Fig. 2.4b we join AB and CD, and as a result CD forces a node at the point where it touches AB. The ground state along AB now has a node and it is a NSDN that introduces an additional phase change of π. In contrast to NSDN discussed by Leggett, this is a physical NSDN and will have physical consequences to be discussed gradually. We will first show that it leads to a deviation from usual parity effect. The phase change for NSDN in the variational wave-function of Leggett does not produce such a deviation. One can identify three different reasons for phase changes in a 1D ring, that are phase changes due to (i) propagation, (ii) Aharonov-Bohm effect, and (iii) anti-symmetry of many body wave-function. Leggett used the NSDN in his variational wave-function to count (in units of π) the phase change due to propagation and said that there will be an integral times π difference between N body wave-function and $N + 1$ body wave-function. The phase change of π seen in Fig. 2.4b is not due to propagation, or Aharonov-Bohm effect or anti-symmetry of wave-functions. It is a new phase change that originates from a physical NSDN.

Let us first try to understand the role of this physical NSDN on parity effect and its consequences. For this we can join the two ends A and B to form a ring as shown in Fig. 2.5. We can now put an Aharonov-Bohm flux through the center of the ring to get a persistent current or magnetization in the ring. The length CD can be varied continuously so that for some specific length of CD we get an exact node forced in the ring as depicted in Fig. 2.4b. For other lengths of CD we will not get an exact node on AB, in which case the states in CD will leak into the ring formed by closing AB. If CD is a few times longer than AB then the states in CD will be more closely

Fig. 2.5 A ring attached to a finite side branch referred to as stub

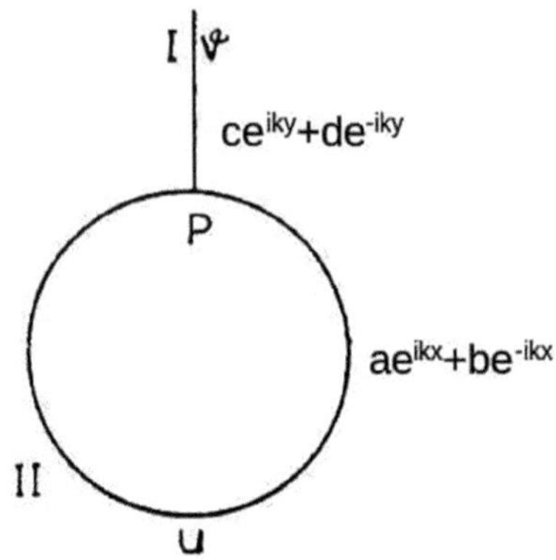

spaced than the states in AB and between two states of isolated AB we will get a few states of CD that leakes into the ring. The question is whether these leaked states will be paramagnetic or diamagnetic?

One can see [2] where we solve for a situation wherein length v of the segment CD, which we now call a stub, can be many times longer than length u of the segment AB which is now formed into a ring (see Fig. 2.5), which means the spacings between the levels of the stub will be much smaller than the levels in the ring. Now when the levels of the stub leak into the ring as they are joined then they will become magnetic and they will show deviations from the usual parity effect. There can be many consecutive states that are all diamagnetic (or paramagnetic). Below we present the calculations and arguments to establish this. Let us solve for the currents in the system shown in Fig. 2.5. The wave-functions in the different regions are also shown in Fig. 2.5. The origin is chosen at the junction point P and the boundary conditions yield the following equations.

$$\begin{aligned} a + be^{-i\alpha} &= ae^{iku+i\alpha} + be^{-iku} = c + d \\ ik(a - be^{-i\alpha}) - ik(ae^{iku+i\alpha} &- be^{-iku}) + ik(c - d) = 0 \\ ce^{ikv} &+ de^{-ikv} = 0 \end{aligned}$$

Here $\alpha = 2\pi\Phi/\Phi_0$ is the AB phase, Φ being the flux through the ring, and Φ_0 is the flux quantum. From these equations one can solve for the allowed modes in the system and is given by the following condition.

$$cos(\alpha) = \frac{sin(ku)cot(kv)}{2} + cos(ku) \tag{2.15}$$

Solving Eq. 2.15 for k at a fixed value of α, we get the allowed wave vectors. One can check that the above condition is the simplified form of the following condition.

$$cos(\alpha) = re(1/T) \tag{2.16}$$

where T is the transmission amplitude across a wire of length u with an attached side arm of length v. This is the structure we get when the ring in Fig. 2.5 is cut open and formed into a 1D line, i.e., we get the system shown in Fig. 2.4b. The above equation immediately suggests the following. For the clean ring the bound state condition is $e^{i(ku-\alpha)} = 1$ (see Eq. 2.3). Whereas Eq. 2.16 is just the condition (It is worth mentioning that $cos(\alpha) = cos(2n\pi - \alpha)$)

$$e^{i[cos^{-1}(re\frac{1}{T})-\alpha]} = 1 \tag{2.17}$$

B$\ddot{u}$ttiker et al. [3] have shown that an electron in a ring with a random potential is effectively moving in a periodic potential whose unit cell is the ring when cut open. It is also well known that $cos^{-1}(re\frac{1}{T})$ is the Bloch phase (Ku where K is the Block momentum) acquired by the electron in traversing an unit cell of an infinite periodic potential where T is the transmission amplitude across the unit cell of the periodic system (see Eq. 8.76 in Solid State physics by Ashcroft and Mermin, International edition 1976). Hence Eq. 2.17 is just $e^{i(Ku-\alpha)}$=1 and is in perfect agreement with Buttiker et al. [3]. It also suggests that inside the ring the electron moves with the momentum K and not with the free particle momentum $\pm k$, although the energy of the state will be $\frac{\hbar^2 k^2}{2m_0}$, and the k values can be obtained from Eq. 2.15. It is to be noted that momentum states are well defined for a electron in an infinite periodic potential as well as for that in a 1D ring, and we have omitted a factor of $\hbar$ for convenience. In a clean ring $(ku - \alpha)$ is the phase acquired by the electron in going round the ring once, whereas in a ring with scatterers (potential or geometric) $(Ku - \alpha)$ is the phase acquired in moving round the ring once and Eq. 2.17 is due to the single valuedness of the wave-function. Inside the ring the electron is moving anticlockwise or clockwise with momentum $\pm K$(where $Ku = cos^{-1}(re\frac{1}{T})$). We will see below that one of them is a diamagnetic (anticlockwise moving) state and the other is a paramagnetic (clockwise moving) state. We will also see that initially as the magnetic flux Φ is increased then the two states ($\pm K$) move away from each other. However, unlike in the case of the clean ring, the diamagnetic state and the paramagnetic state are not degenerate for any value of Φ as the scatterer breaks the rotational symmetry of the ring and lifts the degeneracy between the clockwise and anticlockwise states at $\Phi = n\Phi_0/2$. This degeneracy can be seen for the clean ring as can be seen in Fig. 2.3.

The bound states of the system in Fig. 2.5 can be determined by graphically solving Eq. 2.16 or $re(1/T) = cos(\alpha)$. Therefore, in Fig. 2.6 we show a simple plot where the solid curve is a plot of $y = re(1/T)$ with ku for v/u=0.2. Wherever this curve intersects the straight line $y = cos(\alpha)$, the corresponding k value gives a bound state for the system for a particular value of α. Let us start with α=0 and then $y = cos(0)$ curve is shown in Fig. 2.6 by the dotted straight line. Two consecutive points where the curve $y = re(1/T)$ intersects the straight line $y = cos(0)$ are denoted by A and B in the figure. The corresponding k values are denoted as k_1 and k_2 in the figure. Hence for $\alpha = 0$, k_1 and k_2 will satisfy Eqs. 2.16 and 2.17 and by substituting them in Eq. 2.2 we can get the allowed eigen-energies for the system in Fig. 2.6.

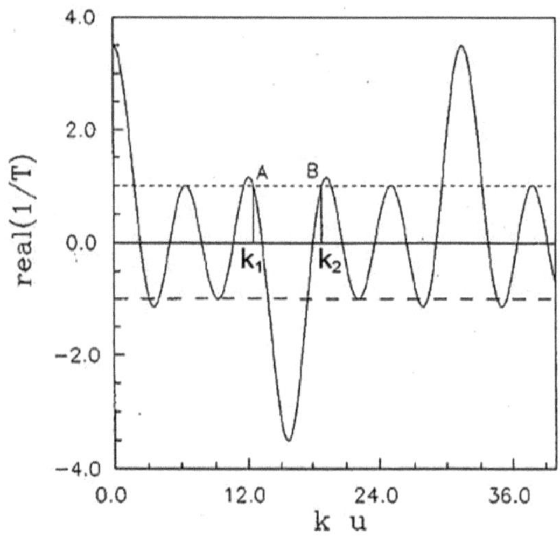

Fig. 2.6 Dispersion curve for the system in Fig. 2.5 for $v/u = 0.2$

If α is increased gradually then the dotted straight line $y = cos(\alpha)$ shifts gradually downwards towards the dashed straight line. As the curve $y = cos(\alpha)$ gradually go downwards the allowed wave vectors k_1 and k_2 slowly drift rightwards and leftwards, respectively, along the k axis. As k_1 drifts rightwards with α, i.e., towards higher energy, k_1 is a diamagnetic state. Similarly k_2 is a paramagnetic state. That k_1, k_2, etc., gradually increase or decrease with α gives rise to a dispersion with α (E vs α) with close by alternate states going further away from each other with α up to $\alpha = \pi$. Note that $y = cos(\pi)$ is also shown in Fig. 2.6 as a dashed line. If we increase α further then the straight curve $y = cos(\alpha)$ starts moving upwards and comes back to its original position at $\alpha = 2\pi$. This ensures Φ_0 periodicity of the dispersion curves. Since $cos(\alpha)$ can vary from -1 to +1 (dotted line to dashed line) the dispersion curve for any two consecutive states can never cross (see Fig. 2.6). So the dispersion curve is exactly similar to that of a ring with a random potential. The cause of gaps in the dispersion curve in that case is the breakdown of rotational symmetry of the ring by the random potential and hence the removal of degeneracy of states that cross for a clean ring. In our case the rotational symmetry is destroyed by the topological defect as compared to potential scattering considered by Buttiker et al. [3]. Some gaps (the ones around $ku = 2n\pi$) are very large but most gaps are very small. Hence from Fig. 2.6 it is evident that alternate states carry persistent currents with opposite signs and have opposite magnetic properties up to infinite energy. This is exactly the same as in case of potential scattering.

The picture is different when we plot the same curves for different values of $v/u = 0.21$ (see Fig. 2.7) (in fact $v/u = 0.2 \pm \epsilon$ is sufficient). Consider the intersections between the curves $y = re(1/T)$ and $y = cos(0)$. The first few alternate states have opposite magnetic properties but the fifth and the sixth states (two consecutive states marked A and B) are both diamagnetic, deviating from the parity effect. One can slowly increase α to see that. Parity effect is again violated for the 11th and the

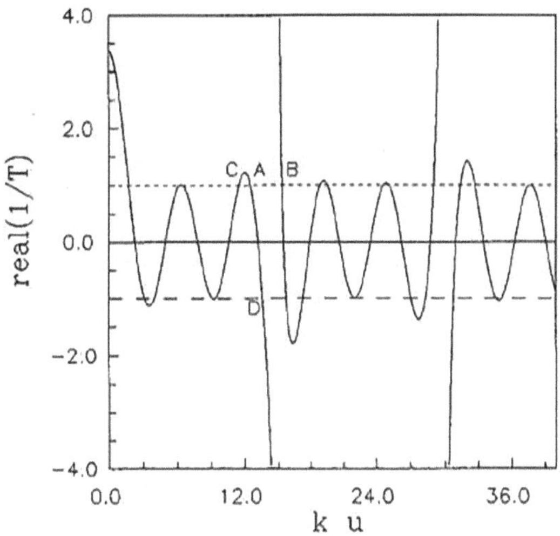

Fig. 2.7 Dispersion curve for the system in Fig. 2.5 for $v/u = 0.21$

12th states both of which are paramagnetic. After a regular spacing of five levels we always find two consecutive levels that violate the parity effect. We shall soon see why it does not happen for specific values of $v/u = 0.2$ (say).

A special feature of the delta potential is that $\mid T \mid^2 = re(T)$. This feature is also observed in the case of a stub. This makes it possible to map a single stub to an effective delta function potential $U(x) = k\ Cot(kv)\delta(x)$. Therefore, the strength of the delta potential depends on k or the Fermi wave vector. That is why at certain energies $U(x)$ become zero and then the gaps vanish. For $k = 0$, we get $U(x) = 1/v$, which means that as k is gradually increased from 0, the effective potential starts with a small positive value. Then it decreases and soon goes to zero. As k is increased further, the strength of the potential monotonously increases on the negative side and finally becomes -∞ at $kv = \pi$. As k is increased further, $U(x)$ undergo a discontinuous jump from $-\infty$ to $+\infty$. If the strengths of the δ potential at $kv = \pi - \epsilon$ and $kv = \pi + \epsilon$ are discontinuous the scattering phase shift and hence the Block phase will also undergo discontinuous jump. Thus, $re(1/T)$ also make a discontinuous jump and hence the Block phase jumps by π (see Fig. 2.7) as the wave vector k or energy is increased (Block phase of the infinite periodic system has to be defined to a modulo of 2π, i.e., $-1 < re(1/T) < 1$). As k is increased, the next possible Block phase of the infinite periodic system of stubs after that at D is that at B and they differ by π. This is markedly different from the next possible Block phase at any other gap, e.g., the Block phase at C is same as that at A. We have seen that if the Block phase of the periodic system of stubs equals the AB phase α (see Eq. 2.16. For the time being we have taken α=0) then the single valuedness of the wave function gets satisfied in the ring and we get a bound state. This additional discontinuous π phase results in satisfying this condition and creating a state at B close to the value $kv = \pi$ which otherwise would not have been there had the phase change across $kv = \pi$ been continuous.

Scattering amplitude of a stub (a finite 1D line) is characterized by a zero-pole pair [4] and such a zero-pole pair is a characteristic feature of Fano resonance. Fano resonance is a fairly well known phenomenon in the case of scattering of light by atoms and molecules, but the discontinuous phase change of π at the zeroes was never discussed and no manifestations of it have ever received any attention prior to our works [2]. Although discontinuous π phase shifts have been discussed earlier for classical waves [5], their role in determining mesoscopic quantum mechanical electronic properties is far reaching as we will show in subsequent chapters. Apart from the breakdown of parity effect it has many dramatic manifestations that will be discussed in subsequent chapters. Fano resonance in low dimensional mesoscopic systems can occur for very natural reasons in a wide range of realistic parameter values [6]. Let us try to understand this further with the help of a schematic diagram that can be modeled and solved, as will be done in Chap. 4. In Chaps. 5 and 6 we will generalize these results and make them independent of any model.

2.3 A Realistic Mesoscopic Grand Canonical System

The system is shown in Fig. 2.8 and it is this general system that we intend to understand through models and schemes in subsequent chapters. This is a more realistic model corresponding to the ideal situation in Fig. 2.4b. The shaded region is the mesoscopic sample and it is connected to a source reservoir and a drain reservoir. The source and the drain have chemical potentials μ_1 and μ_2 as well as temperatures T_1 and T_2, respectively, as shown in the figure. However, these parameters along with the Hamiltonian of the sample are not enough to tell us anything about the properties of the system. We also need to consider the leads L and R that connect the sample to the reservoirs. The exchange of energy and particles between the system and the reservoirs takes place through the leads and the nature of the leads along with the points where it connects to the sample are very important and are well accounted for in the Landauer-Buttiker formalism. If the leads are connected to different points then the properties of the system will change completely. If the confinement potentials of the leads are different then the nature of the states in the leads will be different and again that will change the properties of the system. This is the reason why intrinsic length scales of the sample (shaded region) are not dominant and does not lead to material specific parameters like resistivity. The confinement potentials in the leads act in the transverse direction while in the longitudinal direction from source to drain, the states are plane wave propagating states. In the figure the transverse confinement is taken to be hard wall which makes the transverse states like the states in a 1D infinite well (shown in the leads in Fig. 2.8). These will be mathematically shown in Chap. 4, and here we just provide some motivation based on physical arguments. A few such transverse states are shown in the left lead L as well as in the right lead R. If the transverse confinement is taken to be harmonic then these states will be different and properties of the system will change. The two leads are taken to be identical and in a real situation one lead may be thinner than the other. Again that

Fig. 2.8 Schematic diagram of a general mesoscopic system

will change the system properties drastically. One can connect the system to more than or less than two reservoirs as is the case for making a four probe measurement versus two probe measurement. In simple words the leads are an integral part of the system and have to be accounted for explicitly in terms of the positions where they are attached, exact number of them that are attached, nature of the transverse states in them, etc. When μ_1 and μ_2 are different then we get a non-equilibrium situation and a transport current will flow through the system. When they are the same the system will have typical equilibrium thermodynamic properties. Same holds for T_1 and T_2. The thermodynamic and transport properties of such a grand canonical system can only be understood from the scattering matrix elements for electrons being incident from the source reservoir. These electrons can be transmitted to the drain reservoir or reflected back to the source reservoir. The transverse mode of incidence and the transverse mode to which it is scattered crucially determine the scattering amplitudes. Depending on the experimental situation we have to consider the particular modes involved and an averaging over the modes is not a freedom that we have. In the linear response regime the same scattering matrix that determines the thermodynamic properties also determines the transport properties but the relevant matrix elements vary. In a situation where the sample dimension is less than the inelastic mean free path, only elastic scattering has to be considered. Otherwise one has to use a suitable model to account for inelastic scattering. But the crucial matter of importance is that only these scattering matrix elements include the effect of the nature of modes in the leads and the nature in which the leads connect to the system. Simply the Hamiltonian of the system or its free energy does not include these effects that are the most important to understand the mesoscopic properties.

2.4 Flow Chart of Logic

We present below a flow chart of logic for the readers that can help them understand this book better. At a technical level, this book will be consistent with this flow of logic. Let us start by discussing some crucial aspects of this flow chart shown in Fig. 2.9. First of all, in the von-Neumann picture, it is very important to characterize the functional space which in this case is called the Hilbert space. Let us explain this with an analogy. For example, the symbol "3" is a meaningless figure. What gives it a meaning is the following. We have to first specify the full set of which this symbolic entity is an element and we have to specify inter-relations between the different elements of this full set. It is these inter-relations that give a meaning to this symbol "3" and only after it acquires a meaning, the symbol can be meaningfully used. These inter-relations are stated in the form of some laws. For example, "3" as an element of the set $\mathbb{N}$ or "3" as an element of the set $\mathbb{R}$ are completely different entities. They follow similar laws like associative, commutative, distributive, etc., based on the allowed operations between set elements like addition and multiplication, with subtle differences that make one set a ring while the other a field. In case of quantum mechanics, it is a space instead of a ring or a field and called the Hilbert space, which again means that it is not good enough to say that the wave-function of a system is ψ unless we make it clear that it is an element of the functional space called the Hilbert space. A space requires some notion similar to a metric. The Hilbert space can be represented as $(\mathcal{H}, +, \cdot, \langle \cdots | \cdots \rangle)$. The $\mathcal{H}$ is a set of elements ψ. Elements of this set are related to each other by addition + and a complex-multiplication $\cdot$ that lead to certain laws. The $\langle \cdots | \cdots \rangle$ is called the inner product and that plays the role of a metric that makes $(\mathcal{H}, +, \cdot, \langle \cdots | \cdots \rangle)$ a sesquilinear vector space. This sesquilinear vector space has some subtle aspects that will be further explained in Fig. 2.10. Thus states under the barrier satisfy the Schrodinger equation but do not satisfy many of these subtle aspects and cannot be admitted as elements of the Hilbert space. Thus while quantum tunneling is a reality that deals with particles before they go under the barrier or after they come out of the barrier, quantum state under the barrier is a problem that has remained unsolved since the inception of quantum mechanics. The final reality that we deal with in quantum mechanics is that there is a wave-packet that has a time evolution. This wave-packet can be written as a Riemann integration of wavelets that are plane waves. The wave vector k of these plane waves form a continuum and we need proper measure theory to evaluate density of states which is the number of allowed k values in a certain energy range. Thus these k states are uncountable and do not form the basis states of the Hilbert space. The basis states of the Hilbert space are countable and infinite in number, which essentially means that the basis states have to be indexed by natural numbers, while the allowed k states have to be indexed by real numbers. For example, in Eqs. 4.24 and 4.28, the $\psi(x, y)$ is a function of the many k_m and each of them can vary continuously, there being a sum over m. A particular term in the sum, that is a particular k_m in the leads is in the continuum and it can be matched to the internal wave-function given in Eq. 4.32, where the $\zeta_n^{e/o}$ are the basis states that are countably infinite. Thus in Eq. 4.43 for example, we get an infinite sum on the RHS that involves all the c_n, there being no sum on LHS which has a continuously varying $k_{m''}$. This is very similar to the concept that an irrational number like $\frac{1}{\sqrt{2}}$ can be written as an infinite series involving rational numbers. The $\psi(x, y)$ is a component of a wave-packet in the lead and the wave-packet is a Hilbert space element at any given time. In Fig. 2.10 we symbolically represent the basis states as K_1, K_2, etc. The Hilbert space has to be

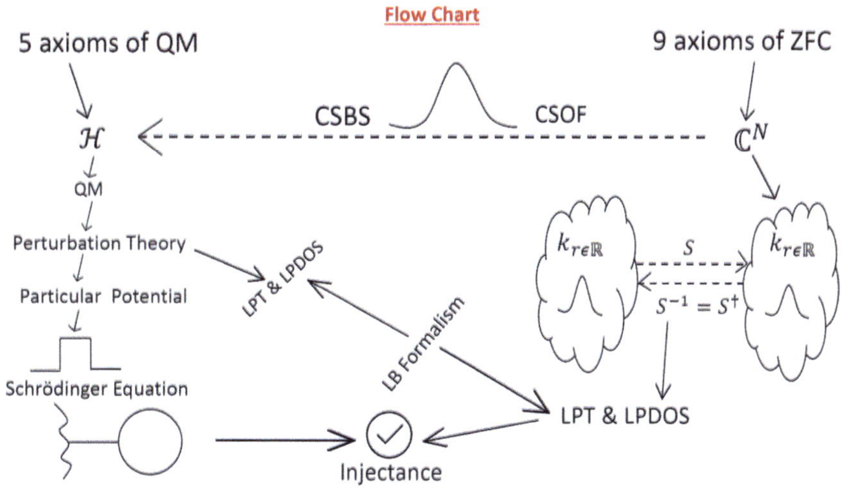

Fig. 2.9 A flow chart of logic

separable, which means the basis states are supposed to be orthonormal by Lebesgue integration. Tunneling states, even though may satisfy the Schrodinger equation, are hyperbolic functions and primarily fail to satisfy these subtle aspects and cannot be admitted as Hilbert space elements. In the case of an infinite periodic potential, the K_1, K_2 states are Bloch states that again do not form a Hilbert space but do form a very similar functional space that is termed the rigged Hilbert space. The problem with hyperbolic functions remains an open problem till date. The completeness of Hilbert space or the rigged Hilbert space implies that although the wave vector k in Fig. 2.10 forms a continuous spectrum, the wavelet for any k value is a Hilbert space element which means the wavelet $a(k)e^{ikx}$ is a point in Hilbert space and this point can be reached through a linear combination of the basis states that are countable and infinite. In analogy consider the fact that our 3D space is continuous and any point in 3D space can be reached by a linear combination of three perpendicular basis states.

In Fig. 2.9 we see reference to two axiomatic frameworks at the top. One is the 5 axioms of quantum mechanics and the other is the 9 axioms (or rather 8 axioms and one postulate) of Zermelo-Fraenkel-choice (ZFC) set theory. The ZFC set theory gives us the number systems that lead to complex numbers and functions which we need for any theory. Quantum mechanics is no exception to that and at some point it imports (a map that involves construction of a wave-packet as explained above) a complete set of orthonormal functions (CSOF) and assign them the role of complete set of basis states. In Chap. 3 we will show that one can take some inputs from perturbation theory of quantum mechanics, idea of spin in quantum mechanics, Larmor precession in classical mechanics, and some inputs from the Landauer-Buttiker formalism (similar to the postulate of random phase approximation of statistical mechanics) to derive the idea of Larmor precession time (LPT) and local partial

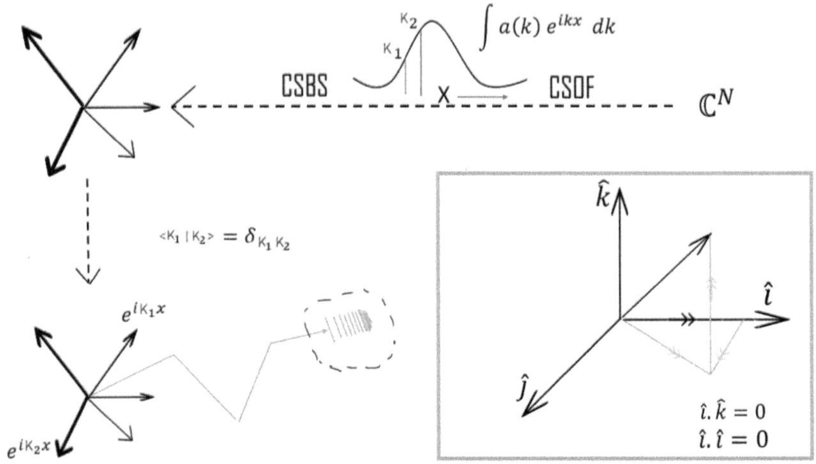

Fig. 2.10 Certain aspects of the Hilbert space

density of states (LPDOS). One cannot get LPT and LPDOS within the axiomatic framework of quantum mechanics. As we go beyond Chap. 3, we will see that one can bypass quantum mechanics, continue with functional analysis, and with some inputs from Landauer-Buttiker formalism one can get the same LPT and LPDOS. We therefore see this as an experimental confirmation of LPT and LPDOS that the axiomatic framework of quantum mechanics denies. It is generally believed that the universe follows the language of numbers and functions. We call this a theoretical experiment and we want to emphasize that a practical experiment is prone to human error and instrumental error. In a theoretical experiment there can be no error apart from careless mistakes that one can easily find and correct.

It is to be noted that for quantum mechanics the functional space is the Hilbert space that explains atomic and molecular spectrum, for solid state physics the functional space is the rigged Hilbert space that explains properties of crystalline materials and thereby also non-crystalline forms of them, in case of condensed matter physics the functional space is Fock space that explains many body effects, but the Schrodinger map leading to the Schrodinger equation is always the same. This signifies that what we call nature is an objective reality, the functional spaces are all artificially constructed but it is the map that connects our artificial theories to nature and we always get the same map, at least for these three different theories. The bigger question remains as to how we can make this map consistent with our classical theories. The scattering matrix S in Fig. 2.9 is also a map that takes the initial or incoming k values to the final or outgoing k values. Now in the von-Neumann approach one tends to say that the Schrodinger equation is fundamental and the scattering matrix is derived from it, this book will present the opposite view that the scattering matrix being a complex function can be classified into some universality classes, and although it appears that we can derive it from the Schrodinger equation, in reality it is determined by the topology of the complex plane which is the common ground between Hilbert space or rigged Hilbert space or Fock space. We can provide a few evidence of that. First is that there are local entities like LPDOS and LPT that can be determined from the scattering matrix but cannot be determined from the Schrodinger equation. That these local objects are real can be clearly seen in presence of NSDN or in presence of Fano resonances. Deviations in other situations can be understood as well. Secondly, as we will show in Chap. 6, that for incident energy tending to zero, the Schrodinger equation fails to give the states correctly or sensibly, while the states determined from scattering matrix do make sense and is consistent with the topology of the complex plane. In the von-Neumann picture the scattering matrix maps between asymptotic states, while in the Landauer picture the reservoirs are an inseparable part of the theory and so the open system with a scattering matrix is fundamental. States of a closed system are a special case of it. The reservoirs are classical and can have its clocks completely synchronized according to the theories of relativity. Thus the quantum theories that follow from the scattering matrix can co-exist with theories of relativity and do not necessarily contradict each other. In the following we take a 1D version of the system shown in Fig. 2.8 and present the original derivation of Landauer conductance formula to make this clear. The current has been derived in chapter one using axioms as explained there and it shows how for a system sandwiched between two reservoirs is given by a scattering matrix element, essentially the transmission amplitude.

Our theoretical experiment uses the analyticity property of the scattering matrix elements. We view the scattering matrix as a map that maps the initial k values of a wave-packet to the final k values of the final wave-packets. Now one may think that we need quantum mechanics to find the scattering matrix which is true. But certain universal properties of the scattering matrix are determined by the topology of the complex plane of the scattering matrix elements, and this book will only use these facts as will be described in detail. Thus LPT and LPDOS are completely consistent with injectance that we can also get from quantum mechanics. Also the analytic principles that establish LPT make it clear that it can become negative and implies that a signal can propagate in negative time. We can use the same analytic properties of the scattering matrix to find current carried by states under the barrier and that is again consistent with the vague ideas that we get from quantum mechanics.

2.5 Landauer Conductance

We have shown in Chap. 1 that the current through the system at a given energy E is $\frac{e_0}{h}|t(E)|^2$ where $t(E)$ is the quantum mechanical transmission amplitude for the square barrier. In Fig. 2.11 we further describe aspects of a mesoscopic system that will help us include finite temperature effects, voltage differences, and role of multiple leads. Our approach in this section will be to provide the original arguments of Landauer and Buttiker and as said before all the conductance formulas given below are experimentally shown to be beyond doubts [see S. Datta, 1995]. If the

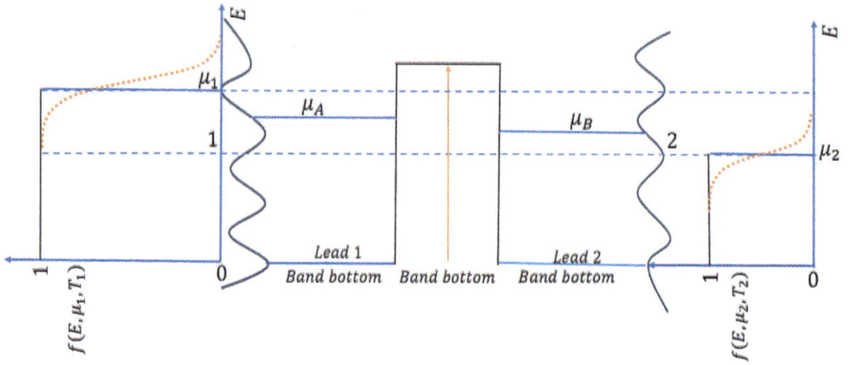

Fig. 2.11 A 1D scatterer is sandwiched between two reservoirs. The source reservoir is to the left and the sink reservoir is to the right of the two wavy lines in the figure. The scatterer in between is shown in the form of a finite square barrier (see Fig. 1.1), whose transmission amplitude is discussed in almost all text books of quantum mechanics. The horizontal axis is the 1D coordinate axis (say x-axis) and the vertical axis is the energy axis. The origin of the energy axis is the band bottom of the conduction band making the sample and the height of the square barrier is shown by an arrow. The 1D square barrier can be practically made from the GaAs-AlGaAs interface by creating a narrow confinement potential in the lateral y and z directions, as will be further discussed in Chap. 4. The reservoirs are classical, have a statistically large number of electrons and the system in between has only quantum states that can transmit electrons from the left reservoir to the right. So the electrons in the reservoirs have a Fermi distribution that at zero temperature resembles a step function while at finite temperature has the well known shape that resembles the dotted lines. The two reservoirs can have different chemical potentials and different temperatures that can be seen from the Fermi distributions shown inside the two reservoirs

square barrier is replaced by any other scatterer, then the transmission amplitude will be different and has to be calculated by solving the scattering problem. At zero temperature, all states in the conduction band are filled below μ_2 and are also equilibrated. Between μ_1 and μ_2, that is in the energy range $(\mu_1 - \mu_2)$, the states can carry a measurable current

$$\frac{e_0}{h}|t(E)|^2(\mu_1 - \mu_2) \tag{2.18}$$

from the left to the right. So in a typical two probe conductance measurement wherein the voltage difference is measured between the two classical reservoirs, the conductance will be

$$G^{zero-temperature}_{two-probe} = \frac{\frac{e_0}{h}|t(E)|^2(\mu_1 - \mu_2)}{\frac{\mu_1-\mu_2}{e_0}} = \frac{e_0^2}{h}|t(E)|^2 \tag{2.19}$$

Now let us see what may happen at finite temperature. The left lead will be in equilibrium with the left reservoir. There are $\frac{dn}{dE}$ number of states in the left lead and not all of them are occupied. Number of electrons in the left lead will be given by the number of occupied states, that is $\frac{dn}{dE} f(E, \mu_1)$ where $f(E, \mu_1)$ is the Fermi distribution function. Similarly, number of electrons in right lead is $\frac{dn}{dE} f(E, \mu_2)$. So number of candidate electrons on the left that wants to cross to the right is

$$\frac{dn}{dE}\Big(f(E, \mu_1) - f(E, \mu_2)\Big) \tag{2.20}$$

Number of them that actually cross to the right is

$$n = \frac{dn}{dE}\Big(f(E, \mu_1) - f(E, \mu_2)\Big)|t(E)|^2$$

Current ne_0v through the system will therefore be

$$\frac{dn}{dE}\Big(f(E, \mu_1) - f(E, \mu_2)\Big)|t(E)|^2 e_0 v \tag{2.21}$$

$$= \frac{1}{hv}\Big(f(E, \mu_1) - f(E, \mu_2)\Big)|t(E)|^2 e_0 v$$

$$= \frac{e_0}{h}\Big(f(E, \mu_1) - f(E, \mu_2)\Big)|t(E)|^2$$

$$= \frac{e_0}{h}\frac{\partial f}{\partial \mu} d\mu |t(E)|^2 = \frac{e_0}{h}(-\frac{\partial f}{\partial E})(\mu_1 - \mu_2)|t(E)|^2 \text{ as } \frac{\partial f}{\partial \mu} = -\frac{\partial f}{\partial E}$$

Total current at finite temperature will be

$$I^{finite-temperature}_{two-probe} = \int_0^\infty \frac{e_0}{h}\Big(-\frac{\partial f}{\partial E}\Big)|t(E)|^2(\mu_1 - \mu_2)dE \tag{2.22}$$

$$I^{finite-temperature}_{two-probe} = \frac{e_0}{h}(\mu_1 - \mu_2)\int_0^\infty dE(-\frac{\partial f}{\partial E})|t(E)|^2$$

For the two probe conductance one has to measure the voltage difference between the two classical reservoirs, which again turns out to be

$$G^{finite-temperature}_{two-probe} = \frac{I^{finite-temperature}_{two-probe}}{\frac{\mu_1 - \mu_2}{e_0}} = \frac{e_0^2}{h}\int_0^\infty dE\Big(-\frac{\partial f}{\partial E}\Big)|t(E)|^2$$

Let us take the zero temperature limit wherein $f(E) \to \theta(E_f - E)$ and $(-\frac{\partial f}{\partial E}) \to \delta(E - E_f)$ to get

$$G^{zero-temperature}_{two-probe} = \frac{2e_0^2}{h}|t(E_f)|^2 \tag{2.23}$$

Since the Fermi energy or the incident energy is determined by the left reservoir we may write

$$G^{zero-temperature}_{two-probe} = \frac{2e_0^2}{h}|t(E)|^2 \tag{2.24}$$

which is Eq. 2.18 which shows that only the density of states of the left lead matters.

For the two probe conductance, the voltage difference between the two classical reservoirs is considered. For a four probe conductance we need the voltage difference between the two leads and the leads are quantum in nature, thus leading to complicacies. Theoretically we can always calculate a voltage difference between the two quantum leads, and so we proceed to calculate the chemical potentials μ_A of left lead and μ_B of right lead, at zero temperature. For a completely opaque barrier we will get the situation $\mu_A = \mu_1$ and $\mu_B = \mu_2$. For a barrier that transmits, we will get a situation $\mu_1 > \mu_A > \mu_B > \mu_2$ as shown in Fig. 2.11. First consider the right lead where the number of available states is given by the DOS and the relevant current carrying states are $\frac{dn}{dE}(\mu_1 - \mu_B)$. We only have to consider the right moving $+k$ (say) states. Of all these available states, a state will be occupied only if it receives an electron transmitted across the scatterer from the left reservoir. Therefore the number of occupied states in right lead between μ_B and μ_1 is

$$n^{oc}_{rl} = \frac{dn}{dE}(\mu_1 - \mu_B)|t(E)|^2 \tag{2.25}$$

There will be also some unoccupied states between μ_2 and μ_B that can be found as follows. The total number of available states in this range is

$$2\frac{dn}{dE}(\mu_B - \mu_2)$$

The factor 2 comes because in this range of energy we have to consider right moving $+k$ (say) states as well as left moving $-k$ (say) states. Suppose, imagine a situation when all these states are unoccupied which means some of the $+k$ states will be filled up by electrons transmitted across the scatterer from the left, which means, of the $\frac{dn}{dE}(\mu_B - \mu_2)$ right moving states, the number of unoccupied ones will be

$$\frac{dn}{dE}(\mu_B - \mu_2) - \frac{dn}{dE}(\mu_B - \mu_2)|t|^2$$

All the $-k$ states will be unoccupied because the only way that these states can be occupied is if they receive an electron from the right reservoir and at zero temperature this cannot happen. Any transmitted electron from left or reflected electron from the right can only go to a $+k$ state, which means all the $\frac{dn}{dE}(\mu_B - \mu_2)$ right moving states in this energy range will be unoccupied in the right lead. So the total number of unoccupied states in right lead will be

$$n_{rl}^{un-oc} = 2\frac{dn}{dE}(\mu_B - \mu_2) - \frac{dn}{dE}(\mu_B - \mu_2)|t(E)|^2 \tag{2.26}$$

Let us now consider the left lead. Number of occupied states between μ_A and μ_1 is

$$n_{ll}^{oc} = \frac{dn}{dE}(\mu_1 - \mu_A) + \frac{dn}{dE}(\mu_1 - \mu_A)|r(E)|^2 = \frac{dn}{dE}(\mu_1 - \mu_A)(1 + |r(E)|^2) \tag{2.27}$$

The first term gives the $+k$ states and all of them are occupied. The second term gives the $-k$ states that are occupied only if there is reflection whose probability is $|r(E)|^2$. Number of unoccupied states in left lead between μ_2 and μ_A will be

$$n_{ll}^{un-oc} = 2\frac{dn}{dE}(\mu_A - \mu_2) - \frac{dn}{dE}(\mu_A - \mu_2)(1 + |r(E)|^2) \tag{2.28}$$

The first term gives the total number of states. From them if we subtract the occupied ones then we will get the number of unoccupied states. So the second term can be understood by looking at Eq. 2.27. Charge neutrality requires

$$n_{ll}^{oc} = n_{ll}^{un-oc} \text{ and } n_{rl}^{oc} = n_{rl}^{un-oc} \tag{2.29}$$

Solving one can get

$$\mu_A - \mu_B = (\mu_1 - \mu_2)|r(E)|^2 \tag{2.30}$$

If this voltage difference can be measured then we will get a conductance that can be obtained from Eqs. 2.18 and 2.30

$$G_{four-probe}^{hypothetical} = \frac{\frac{e_0}{h}|t(E)|^2(\mu_1 - \mu_2)}{\frac{\mu_A - \mu_B}{e_0}} = \frac{e_0^2}{h}\frac{|t(E)|^2}{|r(E)|^2} \tag{2.31}$$

It is labeled hypothetical because this conductance is purely theoretical and cannot be measured. That is analogous to eigen-states of an atom that cannot be observed because any attempt to observe will disturb the state. This also tells us why a conductivity of the sample cannot be defined. This also makes it clear that the Hamiltonian of a system cannot tell us the conductance. As to what will be observed or measured, we provide the following analysis given by Buttiker.

Let us say that there are multiple leads attached to a mesoscopic sample that are labeled p, q, etc. Current flowing from lead p to lead q will be

$$I_{qp} = \frac{e_0}{h} |t_{qp}|^2 (\mu_p - \mu_q) \tag{2.32}$$

Since $t_{qp} = t_{pq}$, we may write

$$I_{qp} = \frac{e_0}{h} (|t_{qp}|^2 \mu_p - |t_{pq}|^2 \mu_q) \tag{2.33}$$

$$I_p = \frac{e_0}{h} \sum_q (|t_{qp}|^2 \mu_p - |t_{pq}|^2 \mu_q) \tag{2.34}$$

We can write the electrostatic potential in the pth reservoir as

$$U_p = \frac{\mu_p}{e_0} \tag{2.35}$$

$$I_p = \sum_q (G_{qp} U_p - G_{pq} U_q) \tag{2.36}$$

where

$$G_{pq} = \frac{e_0^2}{h} |t_{pq}|^2 \tag{2.37}$$

If we make $U_q = U_p$ for all q in Eq. 2.36, then we must get $I_p = 0$ which gives the following sum rule

$$\sum_q G_{qp} = \sum_q G_{pq} \tag{2.38}$$

Using Eq. 2.38 in Eq. 2.36 we get

$$I_p = \sum_q G_{pq} (U_p - U_q) \tag{2.39}$$

First of all micro-reversibility will imply the following which is one of the testing grounds of this theory. Normally one may expect that if magnetic field B is reversed then conductance will be the same but this notion is wrong as it originates from the idea that conductance is a property of the system. But instead, conductance depends on how we look at the system. If we reverse the magnetic field then we must also interchange the leads and only then we get micro-reversibility.

$$\left[G_{pq} \right]_{+B} = \left[G_{qp} \right]_{-B} \tag{2.40}$$

In a laboratory set up an observer in front or above the set up may see a magnetic field directed away from him while an observer behind or below the set up will see a magnetic field directed towards him and both will obviously see the same conductance. But the below/behind situation cannot be obtained by physically reversing the magnetic field. It has to be accompanied by a reorientation of the leads to simulate the same situation. This is a very robust equality that has been never violated in an experiment. Secondly if the pth lead wants to be used as a voltage probe then it should not draw any current or I_p=0. This condition yields a solution to Eq. 2.36 which with Eq. 2.38 gives the following U_p which is what would be the voltage at the point where the pth lead attaches to the sample.

$$U_p = \frac{\sum\limits_{q \neq p} G_{pq} U_q}{\sum\limits_{q \neq p} G_{pq}} \tag{2.41}$$

Let us talk of a three probe conductance instead of a four probe conductance because that is very informative and will be used in Chap. 7. This means p and q take values from the set $\{1, 2, 3\}$, which means lead 1 is from the source reservoir and lead 3 is going into the sink reservoir. That also means U_1 and U_3 are applied by an experimentalist while he is trying to measure U_2 which gives the voltage of the floating point to which lead 2 attaches to the sample. One can choose to set $U_3 = 0$ without any loss of generality. Therefore, Eq. 2.39 can be explicitly written as

$$\begin{bmatrix} I_1 \\ I_2 \\ I_3 \end{bmatrix} = \begin{bmatrix} G_{12} + G_{13} & -G_{12} & -G_{13} \\ -G_{21} & G_{21} + G_{23} & -G_{23} \\ -G_{31} & -G_{32} & G_{31} + G_{32} \end{bmatrix} \begin{bmatrix} U_1 \\ U_2 \\ U_3 \end{bmatrix} \tag{2.42}$$

Not all three equations are independent because of Kirchhoff's law that ensures $I_1 + I_2 + I_3$=0 which leaves us with a 2×2 matrix as we can ignore the third equation and hence the third row in Eq. 2.42.

$$\begin{bmatrix} I_1 \\ I_2 \end{bmatrix} = \begin{bmatrix} G_{12} + G_{13} & -G_{12} \\ -G_{21} & G_{21} + G_{23} \end{bmatrix} \begin{bmatrix} U_1 \\ U_2 \end{bmatrix} \tag{2.43}$$

We have to invert this matrix to get resistance.

$$\begin{bmatrix} U_1 \\ U_2 \end{bmatrix} = \begin{bmatrix} G_{12} + G_{13} & -G_{12} \\ -G_{21} & G_{21} + G_{23} \end{bmatrix}^{-1} \begin{bmatrix} I_1 \\ I_2 \end{bmatrix} \tag{2.44}$$

The resistance matrix will be

$$\begin{bmatrix} U_1 \\ U_2 \end{bmatrix} = \begin{bmatrix} R_{11} & R_{12} \\ R_{21} & R_{22} \end{bmatrix} \begin{bmatrix} I_1 \\ I_2 \end{bmatrix} \tag{2.45}$$

From this we have to solve for U_2 subject to the condition that I_2=0 because then lead 2 will act as a voltage probe. Current through the system will be I_1 and this current comes from the source and goes to the sink. According to Kirchhoff's law convention $I_3 = -I_1$ as current leaving the system will have opposite sign as that which enters the system. So by definition the three terminal resistance will be

$$R_{3t} = \left[\frac{U_2}{I_1}\right]_{I_2=0} = R_{21} \tag{2.46}$$

Whereas, if instead we want to use lead 1 as the voltage probe then note that the indices 1 and 2 in Eq. 2.46 are to be interchanged as now the current I_2 flows through the system, but now it gives a completely different quantity. That is

$$\left[\frac{U_1}{I_2}\right]_{I_1=0} = R_{12} \tag{2.47}$$

That is $R_{21} \neq R_{12}$. If we reverse the magnetic field direction, it is also necessary to re-label the leads to see micro-reversibility.

References

1. Kobayashi K, Aikawa H, Katsumoto S, Iye Y (2003) Phys Rev B 68:235304; Kobayashi K, Aikawa H, Sano A, Katsumoto S, Iye Y (2004) Phys Rev B 70:035319
2. Singha Deo P (1996) Phys Rev B 53:15447
3. Buttiker M, Imry Y, Landauer R (1983) Phys Lett 96A:365
4. Sols F et al (1989) J. Appl Phys 66:3892
5. Berry MV (1998) J Mod Opt 45:1845–1858
6. Singha Deo P (1998) Solid St Commun 107:69

Chapter 3
Larmor Clock and Friedel Sum Rule

Abstract We start this chapter by showing that spin in quantum mechanics is a very counter-intuitive entity. While a rotating classical object in a force field undergoes precession, one cannot identify spin in quantum mechanics with any physical rotation. Yet we can draw some analogies between a quantum spin in an eigen-state with physical rotation to arrive at the pivotal formalism that can allow one to theoretically study the transport and thermodynamic properties of the general mesoscopic system described with respect to Fig. 2.8. It will lead to a hierarchy of formulas leading up to the Friedel sum rule that relates density of states to scattering phase shifts. The hierarchy members are not found in the axiomatic framework of quantum mechanics and we need the Landauer-Buttiker formalism to get them. Application of this formalism is relevant to electronic charge related effects like magnetization, polarization, AC response, non-linear response, etc., because that is where we can directly extend the Landauer philosophy.

In this chapter we present a systematic derivation of Friedel sum rule for mesoscopic systems which is quite different from that used in condensed matter physics in the sense that it consists of a hierarchy of formulas rather than a single formula. It eventually leads to a similar relation between DOS and scattering phase shifts, but with important correction terms to the formula presented in condensed matter textbooks [1]. These formulas form the basic foundation for understanding mesoscopic systems. In Chaps. 5 and 6 we will show that the phase lapses discussed in Chap. 2 further complicate the hierarchy of formulas, in fact in a beneficial manner.

3.1 Spin Recapitulation

The Stern-Gerlach experiments give very important clue to develop the theory of quantum mechanics. In the experiment, an electron beam is passed through an inhomogeneous magnetic field directed in the z-direction (say) that results in a splitting of the beam to give two distinct lines on a screen. The mathematical machinery to understand the spins will be the mathematics of vectors, impact of classical forces

P. S. Deo and K. Meena, *Time Travel: A Reality in Mesoscopic Physics*,
https://doi.org/10.1007/978-981-96-1136-2_3

will be transformations of vectors, and only corresponding to a transformation there can be an observable for which we have to find a suitable operator. The Hamiltonian operator corresponding to the potential energy $-\vec{\mu} \cdot \vec{B}$ is responsible for the splitting. One says that the Hamiltonian dynamics happen in a mathematical vector space called the Hilbert space where as long as the actual observation is not made, all possibilities that can be observed remain in linear superposition. To answer what may happen to a particular spin at the detector, we have to introduce something called "collapse of wave-function" in the theory that is independent of the Hamiltonian. Spins polarized predominantly in $+x$-direction, followed by an effort to split in z-direction, disturb the spins in a random manner, defying classical forces and this is modeled by non-commutating terms in the Hamiltonian.

We will further say from the experiments that the electron is a spin half particle described in a two-dimensional complex vector space. That is the possible space of states of a spin half particle is a two-dimensional complex vector space. We will argue below that this is the correct mathematics (a language) that interprets the seemingly uncorrelated facts observed in the Stern-Gerlach experiment and thereby use the language to deduce what else may happen. It is all about figuring out a mathematics and thereby a representation that describes all aspects of the phenomenon observed in Stern-Gerlach experiment. A finite dimensional vector space can be described by linear algebra, operators being represented by Hermitian matrices with eigenvalues and eigen-functions and at no point we need to talk of functional analysis and self-adjoint operators. The term analysis is specially used when continuous transformations are involved, and when we have to deal with such transformations, one can change the language to functional analysis.

Any arbitrary vector in the two-dimensional complex vector space can be written as

$$|\psi\rangle = c_1 \, |\psi_z\rangle_+ + c_2 \, |\psi_z\rangle_- \tag{3.1}$$

where c_1 and c_2 are complex numbers. If c_1 and c_2 is arbitrary then $|\psi\rangle$ on LHS of Eq. 3.1 represents the state of the arbitrary spin in a magnetic field in z-direction. Corresponding to acting with the Hamiltonian operator results in the two lines corresponding to $|\psi_z\rangle_+$ and $|\psi_z\rangle_-$ with probabilities $|c_1|^2 = |c_2|^2$. The operator accounts for the classical forces due to the magnetic field that supposedly makes an observable transformation of $|\psi\rangle$. The two kets $|\psi_z\rangle_+$ and $|\psi_z\rangle_-$ are notations for the two basis vectors that are very abstract and so it is not relevant to search for a meaning for this notation. They will be orthogonal to each other for the following reason. If we block the lower beam after the splitting and pass the upper beam through another identical Stern-Gerlach set up that gives a force in the z-direction to flip the spins then nothing happens and we only get one line at the detector. That means once polarized in the upper beam it has no overlap with the lower beam. Thereby, thinking in terms of perfect polarization, precession due to classical forces is discarded and the idea of orthogonality of vectors is adopted. It is often convenient to use representations for which we say

$$|\psi_z\rangle_+ = \begin{bmatrix} 1 \\ 0 \end{bmatrix} \qquad |\psi_z\rangle_- = \begin{bmatrix} 0 \\ 1 \end{bmatrix} \tag{3.2}$$

One can also represent $\hat{i}$ and $\hat{j}$ of 2D classical vector space by these two column matrices and one can see that $\langle\psi_{z+}|\psi_{z-}\rangle = 0$. Normalization is a choice by which we can say $|c_1|^2 + |c_2|^2 = 1$.

To describe observables we need operators and in fact linear operators in the above mentioned vector space will work. Once we figure out the operators for this particular vector space, it will be called the spinor space, which means the vector space is not fully defined by the basis vectors alone unless we also define the operators that act on it. The basis vectors are abstract entities and operators are again like maps that connect them to nature. It is the operators that tell us how vectors in this space transform. The classical force due to magnetic field can be oriented in three possible independent spatial directions to produce a split in three possible ways. So we need three operators and we know that linear operators themselves can form a vector space. Just like in 3D classical vector space, a vector can be written as a 1×3 column matrix and corresponding to an operation, say rotation, we can find a rotation matrix that is 3×3, these operators will be 2×2. Consistent with this notation and consistent with the observations of the Stern-Gerlach experiment, we can write the three spin operators as

$$\hat{S}_x = \frac{\hbar}{2}\begin{bmatrix} 0 & 1 \\ 1 & 0 \end{bmatrix} \quad \hat{S}_y = \frac{\hbar}{2}\begin{bmatrix} 0 & -i \\ i & 0 \end{bmatrix} \quad \hat{S}_z = \frac{\hbar}{2}\begin{bmatrix} 1 & 0 \\ 0 & -1 \end{bmatrix} \tag{3.3}$$

The Pauli matrices are

$$\hat{\sigma}_x = \begin{bmatrix} 0 & 1 \\ 1 & 0 \end{bmatrix} \quad \hat{\sigma}_y = \begin{bmatrix} 0 & -i \\ i & 0 \end{bmatrix} \quad \hat{\sigma}_z = \begin{bmatrix} 1 & 0 \\ 0 & -1 \end{bmatrix} \tag{3.4}$$

$$\hat{S} = \frac{\hbar}{2}\hat{\sigma} = \frac{\hbar}{2}(\hat{\sigma}_x + \hat{\sigma}_y + \hat{\sigma}_z) = (\hat{S}_x + \hat{S}_y + \hat{S}_z) \tag{3.5}$$

Now there is a big difference between the 2D classical vector space and this 2D vector space of spins called the spinor space. While $\hat{i} \cdot \hat{j} = 0$, making the x-axis and the y-axis independent of each other meaning operations in one direction is independent of the operations in the other direction, there are special commutation relations between the Pauli matrices, implying an operation in one direction will affect the operation in another direction. That will account for the fact that an effort to polarize in z-direction can throw the spins polarized in $+x$ direction into a dizzy as observed in the experiments. Classically a force in $+z$-direction will not affect magnetic moments perfectly polarized in $+x$-direction. And if the polarization is not perfect then classical forces will have some effect meaning the spins will prefer to orient themselves in $+z$ direction instead of complete dizziness which is not seen. From Eq. 3.4 we can show

$$\hat{\sigma}_x \cdot \hat{\sigma}_y = i\hat{\sigma}_z, \quad \hat{\sigma}_y \cdot \hat{\sigma}_z = i\hat{\sigma}_x, \quad \hat{\sigma}_z \cdot \hat{\sigma}_x = i\hat{\sigma}_y \tag{3.6}$$

$$\hat{\sigma}_x\hat{\sigma}_y + \hat{\sigma}_y\hat{\sigma}_x = 0 \tag{3.7}$$

$$\hat{\sigma}_y\hat{\sigma}_z + \hat{\sigma}_z\hat{\sigma}_y = 0 \tag{3.8}$$

$$\hat{\sigma}_z\hat{\sigma}_x + \hat{\sigma}_x\hat{\sigma}_z = 0 \tag{3.9}$$

$$\hat{\sigma}_x \cdot \hat{\sigma}_x = 1, \quad \hat{\sigma}_y \cdot \hat{\sigma}_y = 1, \quad \hat{\sigma}_z \cdot \hat{\sigma}_z = 1 \tag{3.10}$$

First of all, it is these anti-commutation relations and not the specific representations used in Eqs. 3.2 and 3.3 that explain the observations of the Stern-Gerlach experiment. There are other ways of choosing the Pauli matrices that also give the same commutation relations. Secondly, the coefficient $\frac{\hbar}{2}$ in front of the matrices in Eq. 3.3 is an universal constant implying transformations in this spinor space are discrete. In classical vector space these coefficients can vary continuously. Finally the complex multipliers c_1 and c_2 in Eq. 3.1 should be also noted, which makes it a complex vector space, while for classical vector space we use real numbers. Thus the spinor space is very different from classical vector space. Much of it has been inspired by the angular momentum operators in quantum mechanics, i.e., $\hat{L}_x$, $\hat{L}_y$, and $\hat{L}_z$ have very similar commutation relations. There also the observations give discrete values in $\hbar$ rather than $\frac{\hbar}{2}$. However, $\hat{L} = \hat{r} \times \hat{p}$ has a classical correspondence with classical angular momentum that helps us a lot to arrive at their representations. There is no such classical correspondence in case of spin. Although it is called spin, there is no analogy with anything actually spinning around. Spin is like a fundamental lodestone and Maxwell's equation has to account for it by incorporating the fact that there are no magnetic mono-poles. The entries in the Pauli matrices are entirely figured out to explain Stern-Gerlach experiments and no clues were obtained from a rotation matrix in real space generated by real space angular momentum.

From Eq. 3.5

$$\hat{S}^2 = \hat{S} \cdot \hat{S} = \frac{\hbar}{2}\hat{\sigma} \cdot \frac{\hbar}{2}\hat{\sigma} = (\frac{\hbar}{2})^2(\hat{\sigma_x} + \hat{\sigma_y} + \hat{\sigma_z}) \cdot (\hat{\sigma_x} + \hat{\sigma_y} + \hat{\sigma_z}) \tag{3.11}$$

$$= \frac{\hbar^2}{4}(\hat{\sigma}_x^2 + \hat{\sigma}_y^2 + \hat{\sigma^2}_z + \hat{\sigma}_x\hat{\sigma}_y + \hat{\sigma}_x\hat{\sigma}_z + \hat{\sigma}_y\hat{\sigma}_x + \hat{\sigma}_y\hat{\sigma}_z + \hat{\sigma}_z\hat{\sigma}_x + \hat{\sigma}_z\hat{\sigma}_y)$$
$$= \frac{\hbar^2}{4}[1 + 1 + 1 + 0 + 0 + 0] = \frac{3}{4}\hbar^2 = \frac{1}{2}(\frac{1}{2} + 1)\hbar^2 \tag{3.12}$$

This is consistent with

$$\hat{S}^2 = \frac{3}{4}\hbar^2 \begin{bmatrix} 1 & 0 \\ 0 & 1 \end{bmatrix} = \hat{S}_x^2 + \hat{S}_y^2 + \hat{S}_z^2 \tag{3.13}$$

$$\hat{S} = \left(\hbar\sqrt{s(s+1)}\right)\hat{\sigma} \tag{3.14}$$

where $s = \frac{1}{2}$ is a quantum number. $\hat{S}$ above with quantum number s is not to be confused with our notation of scattering matrix S with matrix elements $s_{\alpha\beta}$ that will appear later. Equation 3.13 is very similar to dot product of two classical vectors, although transformations in different spin directions are not independent of each other. In short, Eq. 3.13 can remind one of Pythagoras theorem with 2×2 matrices. Since $\hat{S}$ and $\hat{S}_z$ commute, they have simultaneous eigen-functions and one can be measured without affecting the other. The same cannot be said for $\hat{S}_x$ or $\hat{S}_y$ and $\hat{S}_z$. One can see from Eqs. 3.2 and 3.3

$$\hat{S}_z \left|\psi_z\right\rangle_+ = \frac{\hbar}{2} \left|\psi_z\right\rangle_+ \tag{3.15}$$

$$\hat{S}_z \left|\psi_z\right\rangle_- = -\frac{\hbar}{2} \left|\psi_z\right\rangle_- \tag{3.16}$$

which means that for $s = \frac{1}{2}$, $\hat{S}_z$ can be spin up or spin down with quantum numbers $s_z = \pm\frac{1}{2}$. Thus we have a complete set of eigenvalues that define the following states.

$$\left|\psi_z\right\rangle_+ = \left| s = \frac{1}{2}, s_z = +\frac{1}{2} \right\rangle \tag{3.17}$$

$$\left|\psi_z\right\rangle_- = \left| s = \frac{1}{2}, s_z = -\frac{1}{2} \right\rangle \tag{3.18}$$

As stated above, the operators themselves form a vector space. The three Pauli matrices with the 2×2 identity matrix that one can see in Eq. 3.13 form the basis states of this vector space. An arbitrary vector in the space spanned by $\hat{S}_x$ and $\hat{S}_y$ will be

$$\hat{S}_\pm = \hat{S}_x \pm i\hat{S}_y \tag{3.19}$$

We should have chosen arbitrary multipliers while forming this linear combination to get an arbitrary vector but we do not do so that will be clear soon. Essentially it will be consistent with the fact that the splitting by each Stern-Gerlach apparatus always gives equal proportions. Different coefficients with the two terms in Eq. 3.19 will give splitting in those proportions. Therefore, substituting from Eq. 3.3 into Eq. 3.19, we get

$$\hat{S}_+ = \hbar \begin{bmatrix} 0 & 1 \\ 0 & 0 \end{bmatrix} \tag{3.20}$$

$$\hat{S}_- = \hbar \begin{bmatrix} 0 & 0 \\ 1 & 0 \end{bmatrix} \tag{3.21}$$

$$\hat{S}_x = \frac{1}{2}(\hat{S}_+ + \hat{S}_-) \tag{3.22}$$

$$\hat{S}_y = \frac{i}{2}(\hat{S}_- - \hat{S}_+) \tag{3.23}$$

Given Eq. 3.17, where complete set of eigenvalues is used, the spin wave-function in x-direction or y-direction can be expanded in terms of $|\psi_z\rangle_\pm$ as they span the spinor space or the spin Hilbert space. From Eq. 3.1

$$|\psi_x\rangle_\pm = \frac{1}{\sqrt{2}}\left[\left|s = \frac{1}{2}, s_z = \frac{1}{2}\right\rangle \pm \left|s = \frac{1}{2}, s_z = -\frac{1}{2}\right\rangle\right] = \frac{1}{\sqrt{2}}\begin{bmatrix} 1 \\ \pm 1 \end{bmatrix} \tag{3.24}$$

$$\left|\psi_y\right\rangle_\pm = \frac{1}{\sqrt{2}}\left[\left|s = \frac{1}{2}, s_z = \frac{1}{2}\right\rangle \pm i\left|s = \frac{1}{2}, s_z = -\frac{1}{2}\right\rangle\right] = \frac{1}{\sqrt{2}}\begin{bmatrix} 1 \\ \pm i \end{bmatrix} \tag{3.25}$$

Note that unless we use complex numbers and use an i in Eq. 3.25, we would not have obtained a $\left|\psi_y\right\rangle_\pm$ that are distinct from $|\psi_x\rangle_\pm$ and mutually orthogonal. With only real coefficients we can only get $\left|\psi_y\right\rangle_\pm = |\psi_x\rangle_\pm$. Thus the Stern-Gerlach set up can be oriented in three possible ways in real space while the vector space is 2D. From Eqs. 3.22 and 3.24 we get

$$\hat{S}_x\,|\psi_x\rangle_\pm = \pm\frac{\hbar}{2}\,|\psi_x\rangle_\pm \tag{3.26}$$

Similarly, from Eqs. 3.23 and 3.25 we get

$$\hat{S}_y\,\left|\psi_y\right\rangle_\pm = \pm\frac{\hbar}{2}\,\left|\psi_y\right\rangle_\pm \tag{3.27}$$

This explains the choice of coefficients in the linear combination of Eq. 3.19. Splitting in any direction is always 50-50 implying the fact that the z-direction is not special. One can orient a Stern-Gerlach set up in x-direction or in y-direction and a second set up following it in the z-direction, or vice verse leading to the same conclusions. It is only the simultaneous operations or the sequence of set ups before the final observation that matter. There can be a sequence in real space, and yet it is a simultaneous operation until an observation is made. One can check

$$|\psi_z\rangle_\pm = \frac{1}{\sqrt{2}}\left[\left|s = \frac{1}{2}, s_x = \frac{1}{2}\right\rangle \pm \left|s = \frac{1}{2}, s_x = -\frac{1}{2}\right\rangle\right] \tag{3.28}$$

$$|\psi_z\rangle_\pm = \frac{1}{\sqrt{2}}\left[\left|s = \frac{1}{2}, s_y = \frac{1}{2}\right\rangle \pm i\left|s = \frac{1}{2}, s_y = -\frac{1}{2}\right\rangle\right] \tag{3.29}$$

One can further check that

$$\langle s = \frac{1}{2}, s_z = \frac{1}{2}|s = \frac{1}{2}, s_x = \frac{1}{2}\rangle = \frac{1}{\sqrt{2}} \tag{3.30}$$

$$\langle s = \frac{1}{2}, s_z = -\frac{1}{2}|s = \frac{1}{2}, s_x = \frac{1}{2}\rangle = \frac{1}{\sqrt{2}} \tag{3.31}$$

One can further see that

$$< \hat{S_x}^2 > + < \hat{S_y}^2 > + < \hat{S_z}^2 > = < \hat{S}^2 > = \hbar^2[s(s+1)] \tag{3.32}$$

which is again like Pythagoras theorem.

3.2 Larmor Precession Time

In this section we present a systematic derivation of the hierarchy of formulas based on the classical concept of Larmor precession. It is an elaboration of the derivation presented by Buttiker [2]. There he also presented several formulas that relate several physical observables like AC response, non-linear response, decoherence, etc., based on this hierarchy of formulas, which, however, has not attracted much attention so far due to the several issues mentioned in Chap. 1. We will derive the hierarchy of formulas in this chapter and in the last chapter discuss some mesoscopic response in terms of them. We will further elaborate these formulas in the next few chapters with respect to the phase lapses discussed in Chap. 2 and show that they are mathematically consistent and physically relevant.

We consider an arbitrary potential $U(y)$ for a quantum particle of unit charge propagating in the y-direction. Asymptotically (i.e., $|y| \to \infty$) $U(y) = 0$ and it is suitable to say that the problem is similar to the scattering problem of Fig. 1.1. The scattering problem is explained in more detail in the caption of Fig. 3.1. The incident particle stationary beam has a spin polarized in the x-direction. Equations 3.30 and 3.31 imply that a magnetic field in z-direction can act on a spin perfectly polarized in x-direction, because such a spin has an overlap with spins oriented in $\pm z$-direction. In the absence of magnetic field $\vec{B}$, the scattering matrix element $s_{\alpha\beta}(E, U(y))$ for spin up as well as spin down particles is a function of energy and functional of potential. Here, α and β are the indices for outgoing and incoming asymptotic modes. Now suppose there is a small non-uniform magnetic field $B(y)$ applied in the z-direction in the region where the potential $U(y) \neq 0$. We are interested in the spin dynamics due to this small non-uniform magnetic field $B(y)$ which means $U(y)$ may well be an electrostatic potential but the magnetic field will add an infinitesimal potential on the spin that is attached to an electron charge. The electronic charge will feel the potential given by the dotted line in Fig. 3.1 while a spin attached to the charge will feel a potential that is given by the difference between the solid line and dotted line in Fig. 3.1 (in 1D $\vec{v} \times \vec{B} = 0$, beside $\vec{v}$ which is always in units of the velocity of light is extremely small and so is $\vec{B}$ which is an infinitesimal perturbation wherein higher order terms will be ignored). The magnetic field however creates a potential $\delta U(y)$ on the spin. Conventionally speaking, a rotating charge in a magnetic field creates a magnetic moment $\vec{\mu} = \frac{e_0}{2m_0}\vec{L}$. This is true classically as well as quantum mechanically as the $\hbar$ factor comes from the quantum mechanical calculation of $\hat{L}$. We may use this analogy to say that associated with the spin of an electron there is a magnetic moment operator $\hat{\mu} = \frac{e_0}{2m_0}\hat{S}_z$. However that is not rigorously true and it needs two corrections that have completely different origins, and yet they almost cancel each other. The first correction is that $\hat{L}$ is quantized in units of $\hbar$ while $\hat{S}_z$ is in units of $\frac{\hbar}{2}$. The second correction comes from Dirac equation that gives $\hat{\mu} = \frac{ge_0}{2m_0}\hat{S}_z$, where $g = 2.002319304362$. The fact that this is very close to 2 for an electron is a rather coincidence because that of a proton is 2.67. Conventionally we say when magnetic field is applied, magnetic moments experience a torque because of which they precess about the magnetic field but this is not at all necessary for the analysis to follow. This statement is generally made to draw an analogy between a precessing top and a precessing magnetic moment. If one wants to think of gravity as a force $\vec{F}$ then one tends to say that there is a torque given by $\vec{\tau} = \vec{r} \times \vec{F}$ which makes the top precess but note that this is just an alternate way of thinking because after all gravity is not a force while electro-magnetic interaction is a force. A spinning top on earth is equivalent to a spinning top in an accelerated reference frame and so the angle of precession has to comply with transformation properties of angles. Electric and magnetic fields too have to satisfy similar transformation properties but trying to draw such parallels is not productive now. To analyze the problem quantum mechanically we can discuss everything in an absolute stationary frame of reference as we will get stationary states as shown below. This is like quantum mechanically calculating the radius of a hydrogen atom where we do not discuss any boosted or accelerated reference frame.

Energy of a magnetic moment in a constant magnetic field $\vec{B}$ is $\delta U = -\vec{\mu} \cdot \vec{B}$. We know that $\hat{S}$ and $\hat{S}_z$ have simultaneous eigen-functions. Hence $\hat{H}_S = -\hat{\mu} \cdot \vec{B} = -\frac{ge_0}{2m_0}\hat{S}_z \cdot \vec{B}$, where we will take $g = 2$. For this Hamiltonian for a quantum spin whose eigen-function will be spin half in positive z-direction denoted as $\begin{bmatrix} 1 \\ 0 \end{bmatrix}$ will have an eigenvalue $E_u = -\frac{ge_0B}{2m_0}\frac{\hbar}{2}$. Similarly, there will be an eigen-function $\begin{bmatrix} 0 \\ 1 \end{bmatrix}$ with an eigenvalue $E_d = \frac{ge_0B}{2m_0}\frac{\hbar}{2}$. Therefore the separation between the two eigenvalues can be written as $\hbar\omega = E_d - E_u = \frac{e_0B}{m_0}\hbar$ or in other words $\omega = \frac{e_0B}{m_0}$. Therefore, in this eigen basis one can write $\hat{H}_S = \omega\hat{S}_z = \omega_L\hat{S}_z$ where ω_L is the classical Larmor frequency in analogy with a precessing angular momentum. We have seen that in going from classical angular momentum to angular momentum operator in quantum mechanics, we can just take that the eigenvalues of the angular momentum operator are quantized in units of $\hbar$. While for spin angular momentum there will be a $g\frac{\hbar}{2} = \hbar$. Thus the time evolution operator is $exp[-\frac{i\hat{H}_st}{\hbar}] = exp[-\frac{i\omega_L\hat{S}_zt}{\hbar}]$. Now the eigen-functions $\begin{bmatrix} 1 \\ 0 \end{bmatrix}$ and $\begin{bmatrix} 0 \\ 1 \end{bmatrix}$ can be used as a basis to write down an arbitrary spin state. So if we start with a spin state polarized in $+x$-direction written as a linear combination (see Eq. 3.24)

$$\left| \frac{1}{\sqrt{2}} \begin{bmatrix} 1 \\ 0 \end{bmatrix} + \frac{1}{\sqrt{2}} \begin{bmatrix} 0 \\ 1 \end{bmatrix} \right\rangle \tag{3.33}$$

After time t, it will be

$$\left| \frac{1}{\sqrt{2}} e^{-i\frac{\omega_L}{2}t} \begin{bmatrix} 1 \\ 0 \end{bmatrix} + \frac{1}{\sqrt{2}} e^{i\frac{\omega_L}{2}t} \begin{bmatrix} 0 \\ 1 \end{bmatrix} \right\rangle$$

We can calculate the expectation value of $\hat{S}_y$ after time t to get

$$< \hat{S}_y >=< \frac{\hbar}{2}\hat{\sigma}_y >= \frac{\hbar}{2} Sin(\omega_L t) \tag{3.34}$$

Similarly we can calculate the expectation value of $\hat{S}_x$ after time t to get

$$< \hat{S}_x >=< \frac{\hbar}{2}\hat{\sigma}_x >= \frac{\hbar}{2} Cos(\omega_L t) \tag{3.35}$$

Thus we get stationary oscillations with frequency ω_L which we call Larmor frequency. Note that while for a rotating top the precession angle increases continuously, the quantum problem yields stationary state solutions. So a y dependent magnetic field is taken to subject the spin to a classical force in the y-direction given by $-\frac{\partial}{\partial y}(-\vec{\mu}.\vec{B}(y)) = -\frac{\partial}{\partial y}(\delta U(y))$ and expect that in the y-direction the spin will change with time.

Thus the scattering matrix element in the presence of magnetic field (for small field) can be expanded as

$$s^{\pm}_{\alpha\beta}(E, U(y) \mp \delta U(y)) = s_{\alpha\beta}(E, U(y)) \mp \int dy' \frac{\delta s_{\alpha\beta}(E, U(y'))}{\delta U(y')} \delta U(y') + ... \tag{3.36}$$

Here $\frac{\delta s_{\alpha\beta}(E,U(y'))}{\delta U(y')}$ is a functional derivative. This is a general situation which in a special case for a differential change in potential at the fixed point $y = y'$ can be written as

$$s^{\pm}_{\alpha\beta}(E, U(y') \mp dU(y')) = s_{\alpha\beta}(E, U(y')) \mp \left[\frac{ds_{\alpha\beta}(E, U(y'))}{dU(y')} \right] dU(y')dy' + ... \tag{3.37}$$

For the initial state in lead 1 (left lead in Fig. 3.1) given by Eq. 3.33, the final state in lead 2 (right lead in Fig. 3.1) will be

$$|\psi_2\rangle = \frac{1}{\sqrt{|s^{+}_{21}|^2 + |s^{-}_{21}|^2}} \begin{bmatrix} s^{+}_{21} \\ s^{-}_{21} \end{bmatrix}$$

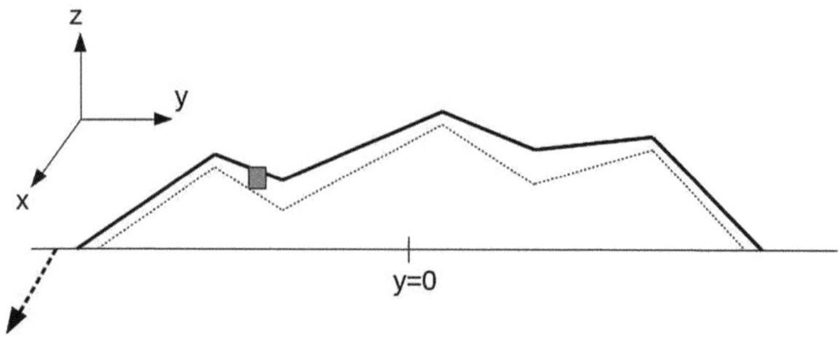

Fig. 3.1 The dashed arrow represents incident particle beam spin polarized in x-direction. Dotted line represents scattering potential $U(y)$. Thick solid line represents potential $U(y) + \delta U(y)$ where $\delta U(y)$ is a very small functional change in potential due to a magnetic field in the z-direction. The magnetic field is non-uniform in the y-direction. A differential change in potential $dU(y')$ at $y = y'$ is shown by the shaded square block between the dotted and solid lines. The area of this block $dU(y')dy'$ is therefore an integration measure. The axis directions and origin marked $y = 0$ are shown in the figure. As usual the region to the left of the potential is the left lead that can be labeled β and that to the right can be labeled α. So the incident particle or spin is coming from β and going towards α

Here the incoming index is $\beta = 1$ and outgoing index is $\alpha = 2$. For example, $\beta = 1$ and $\alpha = 1$ can be used to do the same analysis for reflected particles while $\beta = 2$ sector will correspond to particles incident from the right. Here, $|\psi_2\rangle$ is transmitted wave vector. Here, $s^+_{\alpha,\beta}$ and $s^-_{\alpha,\beta}$ are the scattering matrix elements for spin up and spin down particles. For the scattering defined in Fig. 3.1, the expectation value of the y-component of spin in the transmitted channel is

$$\langle \hat{S}_y \rangle_2 = \langle \psi_2 | \frac{\hbar}{2} \hat{\sigma}_y | \psi_2 \rangle \tag{3.38}$$

The suffix 2 on the LHS means we are looking at the transmitted channel.

$$\langle \hat{S}_y \rangle_2 = \frac{\hbar}{2} \frac{1}{|s^+_{21}|^2 + |s^-_{21}|^2} \begin{bmatrix} s^{+*}_{21} & s^{-*}_{21} \end{bmatrix} \begin{bmatrix} 0 & -i \\ i & 0 \end{bmatrix} \begin{bmatrix} s^+_{21} \\ s^-_{21} \end{bmatrix}$$

$$= \frac{\hbar}{2} \frac{1}{|s^+_{21}|^2 + |s^-_{21}|^2} \left(-i s^{+*}_{21} s^-_{21} + i s^{-*}_{21} s^+_{21} \right) = \frac{-i\hbar}{2} \frac{1}{|s^+_{21}|^2 + |s^-_{21}|^2} \left(s^{+*}_{21} s^-_{21} - s^{-*}_{21} s^+_{21} \right)$$

Using the expansion of $s^{\pm}_{21}$ from Eq. 3.36, we get

$$\langle \hat{S}_y \rangle_2 = \frac{-i\hbar}{2\left(|s^+_{21}|^2 + |s^-_{21}|^2\right)} \left[\left(s^*_{21} - \int dy' \frac{\delta s^*_{21}}{\delta U(y')} \delta U(y') \right) \left(s_{21} + \int dy' \frac{\delta s_{21}}{\delta U(y')} \delta U(y') \right) \right] + \cdots$$
$$+ \frac{i\hbar}{2\left(|s^+_{21}|^2 + |s^-_{21}|^2\right)} \left[\left(s^*_{21} + \int dy' \frac{\delta s^*_{21}}{\delta U(y')} \delta U(y') \right) \left(s_{21} - \int dy' \frac{\delta s_{21}}{\delta U(y')} \delta U(y') \right) \right] + \cdots$$

Multiplying and retaining the terms up to first order in $\delta U(y')$, we get

$$\begin{aligned}\langle \hat{S}_y \rangle_2 &= \frac{-i\hbar}{2T}\left[\int dy' s_{21}^{*}\frac{\delta s_{21}}{\delta U(y')}\delta U(y') - HC\right]\\ &= \frac{-i\hbar}{2T}\left[\int dy'\left(s_{21}^{*}\frac{\delta s_{21}}{\delta U(y')} - HC\right)\delta U(y')\right]\end{aligned}$$

where $2T = |s_{21}^{+}|^2 + |s_{21}^{-}|^2$ and HC stands for Hermitian conjugate. Consider that the magnetic field is constant in a small interval $[y, y + dy]$ and takes there the value B. As $\delta U(y)$ is the perturbation in potential due to the presence of magnetic field, hence $\delta U(y)$ vanishes everywhere except in the interval $[y, y + dy]$, where it takes the value $dU = \frac{\hbar\omega_L}{2}$. The potential is changed by $-\frac{\hbar\omega_L}{2}$ for spin up particles and changed by $+\frac{\hbar\omega_L}{2}$ for spin down particles that leads to the time evolution of a state we get in Eq. 3.34. Therefore, in this interval we can drop the integral and then the y-component of spin is given by the integrand.

$$\Delta\langle \hat{S}_y \rangle_2 = \frac{-i\hbar}{2T}\left[dy'\left(s_{21}^{*}\frac{\delta s_{21}}{\delta U(y')} - HC\right)\left(\frac{-\hbar\omega_L}{2}\right)\right] \tag{3.39}$$

This being a differential change, one can say that for an incident beam with spin polarized in $+x$-direction, the transmitted beam trying to orient as spin up develops a y-component due to a magnetic field at the point y' given by

$$\Delta\langle \hat{S}_y \rangle_2 = \frac{-i\hbar}{2T}\left[\left(s_{21}^{*}\frac{\delta s_{21}}{\delta U(y')} - HC\right)\left(\frac{-\hbar\omega_L}{2}\right)\right] \tag{3.40}$$

For unit spin (note that the circumference of an unit circle is 2π), the y-component developed is given by

$$\frac{\Delta\langle \hat{S}_y \rangle_2}{\frac{\hbar}{2}} = \frac{i\hbar}{2T}\left[\left(s_{21}^{*}\frac{\delta s_{21}}{\delta U(y')} - HC\right)\omega_L\right] \tag{3.41}$$

The y-component of unit spin is equivalent to angle of precession of the particles. Dividing this quantity by Larmor frequency ω_L will give a time which is called Larmor precession time. Essentially in Eq. 3.40 we have calculated $\frac{\hbar}{2}Sin(\omega_L t)$ as we can see by comparing with Eq. 3.34. In absence of magnetic field, that is for the dotted potential in Fig. 3.1, $\langle \hat{S}_y \rangle_2 = 0$ while for an infinitesimal magnetic field at y' $\langle \hat{S}_y \rangle_2 = \frac{\hbar}{2}Sin(\omega_L t)$. So we divide the RHS of Eq. 3.40 by $\frac{\hbar}{2}$ to get Eq. 3.41 for an infinitesimal magnetic field to give $\langle \hat{S}_y \rangle_2 = Sin(\omega_L t) = \omega_L t$. Further if we divide by ω_L then we will get a time called Larmor precession time that will be therefore given by

$$\tau_{21}(y, E) = \frac{i\hbar}{2T}\left[\left(s_{21}^*\frac{\delta s_{21}}{\delta U(y)} - HC\right)\right]$$
$$= \frac{-h}{4\pi i T}\left[\left(s_{21}^*\frac{\delta s_{21}}{\delta U(y)} - HC\right)\right] \tag{3.42}$$

It is to be noted that similarly if we take the leading term in Eq. 3.35 for an infinitesimal magnetic field at y' then we get $\langle \hat{S}_x \rangle_2 \rightarrow \frac{\hbar}{2}$ which is essentially giving the initial spin configuration shown in Fig. 3.1. Note that we are dividing a quantity calculated purely from quantum mechanics by the classical Larmor frequency. This is similar to the Landau level eigen-energy divided by $\hbar$ giving cyclotron frequency $\omega_c = \frac{e_0 B}{m_0}$. In Chap. 5 we will present a rigorous view that this makes perfect sense around a Fano resonance and is a consequence of the fact the scattering matrix elements in quantum mechanics are analytic complex quantities that have to satisfy deeper principles of analyticity (functional analysis being the mother theory for quantum mechanics). In Chap. 6 we will show that the LPT gives the propagation time of an undispersed wave-packet and it can be positive as well as negative. The wave-packet describes a quantum ensemble and Fourier components of the wave-packet define a local time (that is time spent by an electron in a partial state at the point y) that in turn defines a local partial density of states that manifests in local physical effects that can be observed in an experiment to confirm its positivity as well as negativity. An experimentalist can always probe a sample locally, say with an STM tip.

The same calculations can be repeated in quasi-one dimension (which will be described in detail in the next chapter) to get the time spent by an particle going from β to α at point $\mathbf{r}$ within the scattering region as given by

$$\tau_{\alpha\beta}(E, \mathbf{r}) = \frac{-h}{4\pi i |s_{\alpha\beta}(E)|^2} Tr\left(s_{\alpha\beta}(E)^{\dagger}\frac{\delta s_{\alpha\beta}(E)}{\delta U(\mathbf{r})} - s_{\alpha\beta}(E)\frac{\delta s_{\alpha\beta}(E)^{\dagger}}{\delta U(\mathbf{r})}\right) \tag{3.43}$$

Tr is trace which represents summation over incident and outgoing transverse mode (channel) indices which are generally momentum indices while α and β are spatial indices corresponding to the spatial positions where the leads attach to the sample, which means we should always follow different notations for indexing spatial channels (incoming and outgoing) and momentum channels (incoming or outgoing). However, following this convention sometimes unnecessarily complicates notations and will not be used subsequently. So we only need scattering matrix elements and its complex conjugation and no $\dagger$. From now on position vector will be denoted as $\mathbf{r}$. From Larmor precession time, a hierarchy of density of states (DOS) is introduced in Ref. [2] and discussed pedagogically below.

3.3 Hierarchy of Mesoscopic Formulas

So far dU was of the dimension of $\vec{\mu} \cdot \vec{B}$, where $\vec{B}$ is vanishingly small leading to a time $\tau_{\alpha,\gamma}(E, \mathbf{r})$ which gives the time spent by an electron of incident energy E going from lead γ to lead α at the point $\mathbf{r}$ of the potential shown by the dotted line in Fig. 3.1. The vanishingly small dU can be due to an electrostatic potential as well, and to be dimensionally correct $e_0 dU$. And in fact it will be shown subsequently that this notion of time comes from the topological structure of the complex plane and so the exact nature of the perturbation is not important. We now change the notation in order to discuss a hierarchy of relations between scattering phase shifts and DOS to write Larmor precession time of Eq. 3.42 as

$$\tau_{lpt}(E, \alpha, \mathbf{r}, \gamma) = -\frac{h}{4\pi i |s_{\alpha\gamma}(E)|^2}\left[s^*_{\alpha\gamma}\frac{\delta s_{\alpha\gamma}}{e_0\delta U(\mathbf{r})} - \frac{\delta s^*_{\alpha\gamma}}{e_0\delta U(\mathbf{r})}s_{\alpha\gamma}\right] \quad (3.44)$$

$\frac{\delta s_{\alpha\gamma}}{e_0\delta U(\mathbf{r})}$ is a functional derivative of $s_{\alpha\gamma}$ with respect to the local potential $U(\mathbf{r})$ at the point $\mathbf{r}$ inside the sample, E is the incident energy, and e_0 is the electronic charge. The ordering of the arguments on the LHS is important as $s_{\alpha\gamma}$ is a matrix element. Physically it means that an electron going from γ to α spends precisely this amount of time at the point $\mathbf{r}$. Thus all three indices γ, $\mathbf{r}$, and α are spatial indices and their ordering is important in all the subsequent formulas we discuss. Since time spent in a propagation is related to states accessed in the process, both being related to the imaginary part of the Greens function one gets a local partial density of states (LPDOS) ρ_{lpd} defined as $\frac{|s_{\alpha\gamma}(E)|^2}{h}\tau_{lpt}$.

$$\rho_{lpd}(E, \alpha, \mathbf{r}, \gamma) = -\frac{1}{4\pi i}\left[s^*_{\alpha\gamma}\frac{\delta s_{\alpha\gamma}}{e_0\delta U(\mathbf{r})} - \frac{\delta s^*_{\alpha\gamma}}{e_0\delta U(\mathbf{r})}s_{\alpha\gamma}\right] \quad (3.45)$$

The electrons that are involved in going from γ to α are $|s_{\alpha\gamma}(E)|^2$ in number and these being indistinguishable the factor $|s_{\alpha\gamma}(E)|^2$ in Eqs. 3.44 and 3.45 is just an averaging over individual electrons. At zero temperature fermions occupy one state each and for non-interacting fermions doubling the input flux in γ will double the output flux in α as long as we are not in the limit of a completely filled band. We cannot get linear superposition of states in the input channels that was argued in Sect. 1.3 before Eq. 1.36. Numerical simulations [3] suggest that Eq. 3.45 is also valid in presence of electron-electron interaction in the sample. Now we can make an averaging over any one or any two or any three of the coordinates α, $\mathbf{r}$, and γ in Eq. 3.45 and accordingly get higher members of DOS in the hierarchy. Summing over γ means averaging over all incoming channels. Summing over α means averaging over all outgoing channels. Accordingly different members can be physically interpreted. Thereby partial density of states (PDOS) for electrons going from γ to α is defined as

$$\rho_{pd}(E,\alpha,\gamma) = -\frac{1}{4\pi i}\int_\Omega d^3\mathbf{r}\left[s^*_{\alpha\gamma}\frac{\delta s_{\alpha\gamma}}{e_0\delta U(\mathbf{r})} - \frac{\delta s^*_{\alpha\gamma}}{e_0\delta U(\mathbf{r})}s_{\alpha\gamma}\right] \tag{3.46}$$

Here Ω stands for the spatial region of the sample only, excluding the leads, which means the region enclosed by the wavy lines in Fig. 3.2. Injectivity of lead γ to the point $\mathbf{r}$ can be defined as

$$\rho_i(E,\mathbf{r},\gamma) = -\frac{1}{4\pi i}\sum_\alpha\left[s^*_{\alpha\gamma}\frac{\delta s_{\alpha\gamma}}{e_0\delta U(\mathbf{r})} - \frac{\delta s^*_{\alpha\gamma}}{e_0\delta U(\mathbf{r})}s_{\alpha\gamma}\right] \tag{3.47}$$

$s_{\gamma,\gamma}$ will correspond to a reflection amplitude. Emissivity of lead α from the point $\mathbf{r}$ can be defined as

$$\rho_e(E,\alpha,\mathbf{r}) = -\frac{1}{4\pi i}\sum_\gamma\left[s^*_{\alpha\gamma}\frac{\delta s_{\alpha\gamma}}{e_0\delta U(\mathbf{r})} - \frac{\delta s^*_{\alpha\gamma}}{e_0\delta U(\mathbf{r})}s_{\alpha\gamma}\right] \tag{3.48}$$

Injectance of lead γ can be defined as

$$\begin{aligned}\rho(E,\gamma) &= -\frac{1}{4\pi i}\int_\Omega d^3\mathbf{r}\sum_\alpha\left[s^*_{\alpha\gamma}\frac{\delta s_{\alpha\gamma}}{e_0\delta U(\mathbf{r})} - \frac{\delta s^*_{\alpha\gamma}}{e_0\delta U(\mathbf{r})}s_{\alpha\gamma}\right]\\ &= -\frac{1}{2\pi}\int_\Omega d^3\mathbf{r}\sum_\alpha |s_{\alpha\gamma}|^2\frac{\delta\theta_{s_{\alpha\gamma}}}{e_0\delta U(\mathbf{r})}\end{aligned} \tag{3.49}$$

Here $s_{\alpha\gamma} = |s_{\alpha\gamma}|exp(i\theta_{s_{\alpha\gamma}})$. Local density of states (LDOS) at $\mathbf{r}$ can be defined as

$$\rho_{ld}(E,\mathbf{r}) = -\frac{1}{4\pi i}\sum_{\alpha\gamma}\left[s^*_{\alpha\gamma}\frac{\delta s_{\alpha\gamma}}{e_0\delta U(\mathbf{r})} - \frac{\delta s^*_{\alpha\gamma}}{e_0\delta U(\mathbf{r})}s_{\alpha\gamma}\right] \tag{3.50}$$

Finally we get DOS where all that can be averaged is summed to give

$$\rho_d(E) = -\frac{1}{4\pi i}\sum_{\alpha\gamma}\int_{sample} d^3\mathbf{r}\left[s^*_{\alpha\gamma}\frac{\delta s_{\alpha\gamma}}{e_0\delta U(\mathbf{r})} - \frac{\delta s^*_{\alpha\gamma}}{e_0\delta U(\mathbf{r})}s_{\alpha\gamma}\right] \tag{3.51}$$

$$\rho_d(E) = -\frac{1}{2\pi}\sum_{\alpha\gamma}\int_{sample} d^3\mathbf{r}\left[|s_{\alpha\gamma}|^2\frac{\delta\theta_{s_{\alpha\gamma}}}{e_0\delta U(\mathbf{r})}\right] \tag{3.52}$$

It is to be noted that a particular channel α can act as an incoming channel as well as an outgoing channel and so double counting in the sums in Eqs. 3.50, 3.51, and 3.52 is allowed. Equation 3.52 is the mesoscopic version of the Friedel sum rule that relates scattering phase shift to DOS and does not depend on the lead indices or coordinate as they have been summed. But there are several lower members that explicitly depend on the lead indices and let us discuss one of them, say $\rho_i(E, \mathbf{r}, \gamma)$, that is injectivity of lead γ to the remote point $\mathbf{r}$ and others can be similarly interpreted. It depends on the input lead index γ and physically means the following. The quantity applies to only those electrons that are incident along lead γ. Individual members of this ensemble of electrons may or may not pass through a remote point $\mathbf{r}$ and in fact there is no equation of motion that tells us whether it will. Note that Schrodinger equation works only for an ensemble and gives us only a probability for it. At zero temperature below Fermi energy when we do not distinguish between counting electrons (that constitute a current) and counting states, $\rho_i(E, \mathbf{r}, \gamma)$ give the fraction of those electrons that come from γ and pass through $\mathbf{r}$. Quantum mechanics with its probabilistic interpretations is incapable of pointing at an arbitrary electron at point $\mathbf{r}$ and say if it came from lead γ and even if we try to do so it messes up the situation. Yet we can calculate this member of hierarchy which is a count or measure of such electrons and in case of some further lower members one has to specify to which lead the electron finally goes. For example ρ_{lpd} is a measure of electrons going from γ to α condition to the fact that they pass through the point $\mathbf{r}$. This may give the impression that some of these members are over specified purely theoretical entities as there is no equation of motion for answering where does an electron coming from γ go after some time. But such $\mathbf{r}$ dependent DOS can be theoretically derived using the idea of physical clocks like Larmor clock. We will see that such $\mathbf{r}$ dependent DOS can be demonstrated in experiments.

Equation 3.52 is the mesoscopic version of Friedel sum rule that relates scattering phase shift to DOS. In this equation too we get the derivative of the scattering phase shift with respect to the local potential $U(\mathbf{r})$ along with an integration over $\mathbf{r}$ and so this is not of much use to an experimentalist. One may consider the following substitution:

$$\int_{global} d^3\mathbf{r} \frac{\delta}{e_0 \delta U(\mathbf{r})} = -\frac{d}{dE} \tag{3.53}$$

Therefore

$$\int_{sample} d^3\mathbf{r} \frac{\delta}{e_0 \delta U(\mathbf{r})} \approx -\frac{d}{dE} \tag{3.54}$$

Thus

$$\rho_d(E) \approx \frac{1}{2\pi} \sum_{\alpha\beta} \left[|s_{\alpha\beta}|^2 \frac{d\theta_{s_{\alpha\beta}}}{dE} \right] \tag{3.55}$$

This is the FSR that appears in textbooks [1]. This is true because the global integration on LHS of Eq. 3.53 implies that the potential can be increased (or decreased) by a constant amount globally which is identical to not shifting the potential but decreasing (or increasing) the incident energy. An experimentalist can easily achieve this by changing the Fermi energy but what appears in Eq. 3.52 is not a global integration but a sample integration. An infinitesimal constant change in $U(\mathbf{r})$ over a finite region of space is a change $\delta U(\mathbf{r})$. For a large bulk system one need not make a difference between the two but for a mesoscopic system we may only write the approximate equality in Eq. 3.55.

Injectance is the first member that can be defined and calculated in quantum mechanics as it requires us to specify only one incoming channel, all outgoing channels being summed and an integration made over all coordinates inside the sample. This can be calculated canonically or using Feynman paths, identically. Injectance is total injected current at zero temperature when only lead γ brings electrons into the system while all other leads carry electrons away from the system and $\mathbf{r}$ is also integrated out. This situation is shown in Fig. 3.2 as a theoretician will set up the scattering problem to be solved. Injected current is of the form ne_0v or differential current is $\frac{dn}{dE}e_0vdE$. Electronic charge e_0 can be set to unity and if properly normalized wave-functions are taken (see Eq. 1.10) then we can also drop the v factor making injected current to be $\frac{dn}{dE}$ at an energy E. Current through the sample can be determined from internal wave-function $\psi(\mathbf{r},\gamma)$ when the scattering problem is also set up such that electrons are incident only along lead γ and all other leads carry electrons away from the system so that we do not have to bother where goes the electrons that pass through $\mathbf{r}$. That gives

$$\rho(E,\gamma) = \int_{sample} d^3\mathbf{r} \sum_{k_\gamma} |\psi(\mathbf{r},\gamma)|^2 \delta(E - E_{\gamma,k_\gamma}) \tag{3.56}$$

$$= -\frac{1}{2\pi}\int_\Omega d^3\mathbf{r} \sum_\alpha \left[|s_{\alpha\gamma}|^2 \frac{\delta\theta_{s_{\alpha\gamma}}}{e_0\delta U(\mathbf{r})}\right] \tag{3.57}$$

Equation 3.56 is the textbook definition of DOS when only lead γ brings electrons into the system and in lead γ the wave vector is k_γ. Equation 3.57 follows from Eq. 3.49. Lower members of the hierarchy cannot be defined in terms of internal wave-function $|\psi(\mathbf{r},\gamma)|^2$ as one can never write down an internal wave-function that depends on two lead indices and $\mathbf{r}$ (over specification). But they can still be defined in terms of the scattering matrix elements or asymptotic wave-functions far away from $\mathbf{r}$. By appealing to physical process like spin precession and Larmor frequency we can address issues like a particle going from γ to α how much time it spends at the point $\mathbf{r}$ and how many (a count or a measure) partial states it occupied at the point $\mathbf{r}$. We thought wave-function is the most fundamental entity in quantum mechanics that is determined once we know the internal potential $U(\mathbf{r})$ and hence the Hamiltonian. We always thought that a state is an entity in Hilbert space. Local density of states can only be defined through ensemble averaging wherein equal apriory probability implies that all states in Hilbert space are equally accessible by the electrons and time averages give phase space averages. Averaging over all possible variations in $U(\mathbf{r})$ helps taking the problem from Hilbert space to phase space. But say for a benzene molecule attached to leads if we change the internal potential $U(\mathbf{r})$, then it is no longer a benzene molecule. An electron coming from lead γ and going to lead α will not access all states of the benzene molecule but some partial states for which Eq. 3.46 gives partial density of states that cannot be defined in terms of the internal wave-function. The integrand in Eq. 3.56 cannot be broken down into an α dependent quantity. If we remove the integration over $\mathbf{r}$ in Eq. 3.56 then the integrand does not give any lower member of hierarchy as fixing an $\mathbf{r}$ means infinite uncertainty in momentum and it is not enough to be limited to a particular k_γ at energy E. Likewise, removing the sum over k_γ would mean looking at a particular state and that would mean an infinite uncertainty in coordinate of the electron and so integrating the coordinate over the sample region does not give anything. Besides a delta function cannot be written unless there is a sum or an integration over its variable. We can of course remove the spatial integrations in Eqs. 3.56 and 3.57 and get an equality that is very counter-intuitive. Derivations are based on the vague idea of spin precession and Larmor frequency. It is this equality that we will justify in Chap. 5 as a topological property of the complex plane, using BC and AD which yields the fact that ρ_{lpd} play the role of I in Eq. 5.1. Further justifications will be given in Chaps. 6 and 7. We thus get a physical entity that has to be defined with respect to α and γ although they do not appear in Eq. 3.56. The integrand in Eq. 3.56 can mean something physical for bulk systems only if ensemble averaged.

Note that the injectance as defined in Eqs. 3.56 or 3.57 can also be measured by an experimentalist. Just as Landauer resistance of a sample is defined with respect to the leads attached to the sample, the injectance in Eq. 3.56 is a property of the lead γ which is dependent on the sample to which it is attached. Any electron incident from γ at a particular energy E and wave vector k_γ has to go to some point $\mathbf{r}$ in the sample and eventually escape to some lead α including escape to lead γ and we have summed over all these possibilities in Eq. 3.57. Current delivering capacity of lead γ will be determined by the lateral confinement in the lead and the available modes. To measure this capacity of an STM tip (say) one has to just bring it close to a Landauer type reservoir or a perfect sink of electrons and measure how much tunneling current it delivers to the sink which will be independent of incident energy and the point of the sink to which it is attached. But once the STM tip tries to deliver a current to a particular point $\mathbf{r}_p$ of a mesoscopic sample then this delivered current will strongly depend on the point $\mathbf{r}_p$ and incident energy E. This E dependent current is due to the available states in the system that the injected electrons can access and can never be found by box normalization. In Landauer's phenomenology, the DOS in the lead per incident mode is $\frac{1}{hv}$, that is the same as an infinite 1D line obtained from box normalization. We understand this coming from an incident wave-packet made up of wavelets with appropriate normalization constants. A wave-packet is a localized pattern in physical space, propagating in time and does not care if the lead is attached to a sample or to a perfect sink. The particular normalization constant in Eq. 1.10 takes care of current conservation via the available current carrying states. The wavelets satisfy Schrodinger equation with or without the normalization constant. We are thereby using a freedom as we did in Eq. 1.27. In Chap. 7 we will show how some lower members of the hierarchy can also be measured experimentally.

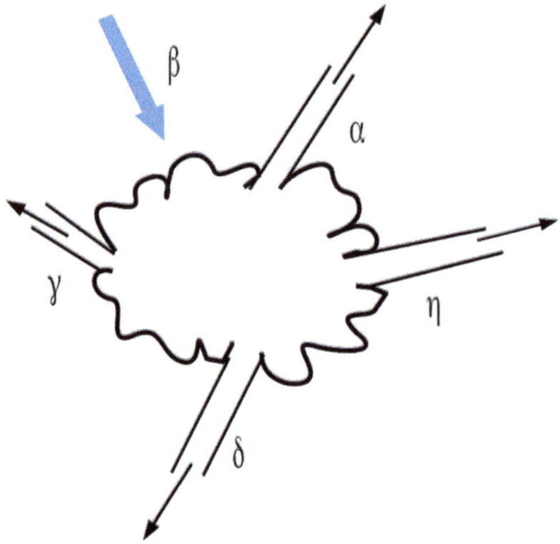

Fig. 3.2 We show a mesoscopic set up where there are many leads α, β, γ, δ, etc., attached to a sample. The sample is the region enclosed by the wavy lines. One can adjust the chemical potential of the leads such that flow of current occurs along the arrow directions shown. In general any lead can bring or take current away from the system. The lead β is special in the sense that it is an STM tip that can deliver (or draw) current to (or from) a variable point **r** in the sample. All other leads are fixed

In 1D α and γ can each take two values leading to the fact that $s_{11} = r$, the reflection amplitude when incidence is from the left. Similarly, $s_{21} = t$, the transmission amplitude when incidence is from the left, $s_{12} = t$, the transmission amplitude when incidence is from the right, and $s_{22} = r^*$, the reflection amplitude when incidence is from the right. There are many other conventions of representing this that does not compromise the physical reality. We can substitute these values in Eq. 3.57 and then use the substitution in Eq. 3.54 to get

$$\rho(E, 1) \approx \frac{1}{2\pi}\left[|r|^2\frac{d\theta_r}{dE} + |t|^2\frac{d\theta_t}{dE}\right] \tag{3.58}$$

Note that when talking of injectance of lead 1, we have to only consider incidence from the left. In 1D it was shown [4] that the correction term can be determined in 1D to give

$$\rho(E, 1) = \frac{1}{2\pi}\left[|r|^2\frac{d\theta_r}{dE} + |t|^2\frac{d\theta_t}{dE} + \frac{m_0|r|}{\hbar k^2}Sin(\theta_r)\right] \tag{3.59}$$

which means the last term is the correction term to the semi-classical formula due to quantum interference effects. If we consider a semi-classical regime where the electron behaves like a classical particle then $Sin(\theta_r)$ goes to zero and we recover the text book FSR. One way to realize the semi-classical regime is to take an ensemble averaging that washes out any dependence on $Sin(\theta_r)$ and make the system effectively behave semi-classically. Talking of a purely quantum system without invoking principles of statistical mechanics like ensemble averaging, one can say that as $|r| \rightarrow 0$, then we can see a system behaving semi-classically. This is known as the WKB regime wherein a quantum particle above the barrier can be completely transmitted if the scale at which the potential varies is much greater than the wavelength of the quantum particle. In Chaps. 4, 5, and 6 we will talk of a situation wherein $|r| \nrightarrow 0$ and yet the correction term goes to zero because $Sin(\theta_r) = 0$. The evanescent modes have to be correctly accounted for to give this in a purely quantum regime and the semi-classical formula becomes exact in a quantum regime. Semi-classical formulas are always easy to deal with, specially in this case, when as we see in Eq. 3.58 we need to make small changes in the incident energy and calculate the derivatives. So if semi-classical formula becomes exact in a quantum regime, it can be of great use to an experimentalist or a theoretician. We will keep referring to Eq. 3.58 as the semi-classical formula although we have expressed and will show that it can become exact in a quantum regime. While Chap. 4 gives a couple of realistic examples that can be solved exactly, Chaps. 5 and 6 will show the generality of the behavior making it practically applicable.

References

1. Ziman JM (1979) Principles of the theory of solids, 2nd edn. Cambridge University Press
2. Buttiker M (2001) Time in quantum mechanics. Springer, pp 279–303
3. Mukherjee S et al (2011) Physica E 44:62
4. Leavens CR, Aers GC (1989) Phys Rev B 39:1202

Chapter 4
Scattering in Q1D

Abstract In this chapter we will work out a few typical cases to calculate the relevant quantities explained in the previous chapter. Namely, the scattering amplitudes for systems of the type explained with Fig. 2.8 and a few members of the hierarchy of DOS for them. The analysis is also an example for von-Neuman approach to quantum mechanics. The new phase shift discussed in Chap. 2 and FSR like rules will also be revealed for these typical cases.

In this chapter we briefly outline how to solve a scattering problem for a realistic mesoscopic system that is based on the work of Bayman and Mehoke [6]. We will present some extra calculations along the same line and point out some new results for these particular systems which will be generalized in the next two chapters from general principles and theorems. These extra calculations were first reported in Singha Deo [1]. We will point out that from these solutions one can verify the phase slips of π that has somehow eluded attention before our works. Its implications will be discussed too. These scattering problems are generally hard to solve and numerically even simple cases require a good amount of computation time [2]. The only potential that can be solved exactly is that of a delta function potential [3] to give the scattering amplitudes. We will briefly outline Bagwell's solution and then proceed to give a detailed derivation of the DOS for the same which is very instructive.

Before going into the mathematical analysis we discuss how the Fano resonances and the associated phase slips appear very naturally in quantum wires. Consider a simplified version of Fig. 2.8 as shown in Fig. 4.1. Here the shaded region of Fig. 2.8 has been simplified into a rectangular block extending from $x = -a$ to $x = a$ and from $y = -c/2$ to $y = c/2$. The leads extend from $y = -b/2$ to $y = b/2$. The mathematical solution of Sc. Eq. for this system will be given in Sect. 4.1. This system can be further simplified by making c equal to b in which case we will get the structure shown in Fig. 4.2c. All the systems in Figs. 2.8, 4.1, and 4.2c will exhibit Fano resonances along with discontinuous phase shifts of π of the nature discussed in Sect. 2.2. The physical reasons can be diagrammatically explained as follows. For example, consider the modes in the leads in Fig. 2.8. The dotted sinusoidal curve is the wave-function for the ground state in an infinite potential well as the potential in the direction transverse to the direction of current flow is taken as hard

P. S. Deo and K. Meena, *Time Travel: A Reality in Mesoscopic Physics*,
https://doi.org/10.1007/978-981-96-1136-2_4

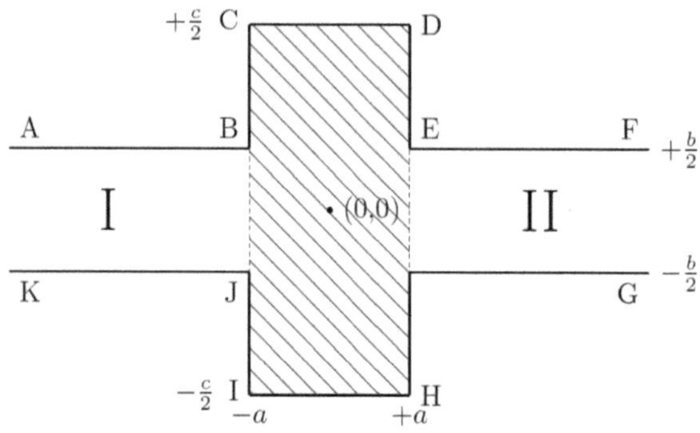

Fig. 4.1 A simplified version of the system in Fig. 2.8

wall. This is just the wave-function of the first state in Fig. 2.1, but now turned vertically. The wave-function in the propagating direction will be plane waves with usual parabolic dispersion curve as we will show in Sect. 4.1. The dotted parabolic curve in Fig. 4.2d is showing this dispersion (it is basically a plot of energy versus wave number). The solid sinusoidal curve in the leads of Fig. 2.8 is the first excited state (depicted as second state in Fig. 2.1) in an infinite potential well and for this we will get another parabolic dispersion curve shown by dashed curve in Fig. 4.2d. This parabola will be at a higher energy relative to the dotted parabola as the first excited state in the transverse direction has higher energy than the ground state (will be shown mathematically in Sect. 4.1). Similarly corresponding to the dashed sinusoidal curve in Fig. 2.8 we will get the solid parabola in Fig. 4.2d. These parabolic dispersion curves correspond to scattering states only. If potential U in the sample region is negative (say $-U_0$) then corresponding to each parabola there will be bound states as well whose energies will be below the threshold energy of the parabolas. Thus the solid parabola will have some bound states shown by the solid straight lines below the solid parabola in Fig. 4.2d. The dashed parabola will also have some bound states shown by the dashed straight lines below the dashed parabola in Fig. 4.2d. Now the dashed bound states will not be true bound states as they are degenerate with a scattering state (shown by an arrow), of the dotted channel. We know that degeneracy between a bound state and a scattering state will always produce a Fano resonance [4] that will be accompanied by a discontinuous phase change of π at the zeroes. However, the mesoscopic picture is rather different for the reservoirs as we will see in subsequent chapters. There will be no true transmission zero but a minimum in the transmission and no discontinuous phase change but instead a sharp and continuous phase drop that will dramatically change our idea about Friedel sum rule like formulas and our understanding of time. Phase drop is a rather novel phenomenon because our intuitive understanding of scattering says that time delays in electron propagation and phase accumulation of the wave-function are related. So phase drop may naively suggest time gains rather than delays. But one has to carefully isolate what is a true physical phenomenon and what is artifact of approximations and this will be done in subsequent chapters.

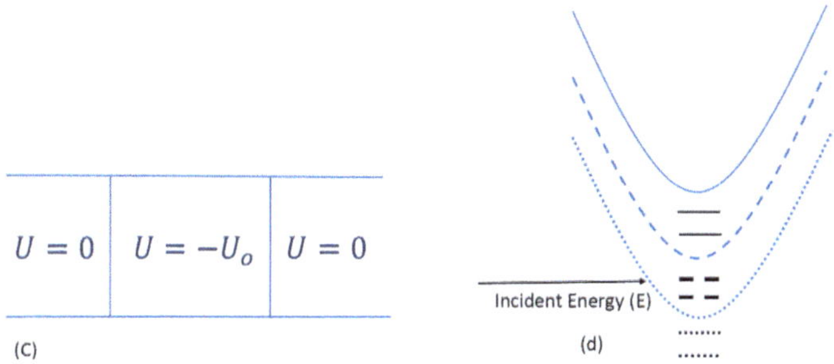

Fig. 4.2 An intuitive picture of getting Fano resonances in mesoscopic systems

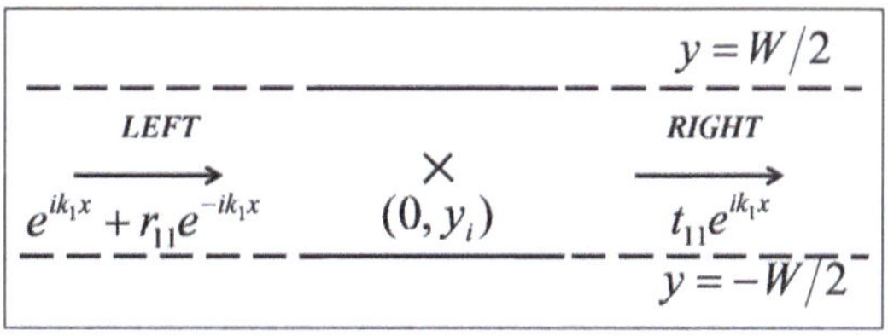

Fig. 4.3 A Dirac delta function potential in a quantum wire

Suppose in Fig. 2.8 or Fig. 4.1 or Fig. 4.2 we want to consider just two transverse channels, which means in case of Fig. 2.8 we have the dotted curve and the solid curve depicting transverse modes. Suppose the propagating wave vectors for these two channels are k_1 and k_2. As Eqs. 3.56 and 3.57 suggest we have to define injectance separately for these two cases. Injectance of left lead for incidence along k_1 channel should be denoted as $\rho(E, \gamma = 1, k_1)$ but we call it $\rho(E, 1)$, which means although according to Eq. 3.43 one has to make a distinction between spatial indices and momentum indices, here we simplify the notation. For this case semi-classical injectance will have two more terms compared to Eq. 3.58 and is given by

$$\rho(E, 1) \approx \frac{1}{2\pi}\left[\mid r_{11} \mid^2 \frac{d\theta_{r_{11}}}{dE} + \mid r_{21} \mid^2 \frac{d\theta_{r_{21}}}{dE} + \mid t_{11} \mid^2 \frac{d\theta_{t_{11}}}{dE} + \mid t_{21} \mid^2 \frac{d\theta_{t_{21}}}{dE} \right] \tag{4.1}$$

We can break this up as

$$\rho(E, 1) \approx \rho^L(E, 1) + \rho^R(E, 1) \tag{4.2}$$

where

$$\rho^L(E, 1) = \frac{1}{2\pi}\left[\mid r_{11} \mid^2 \frac{d\theta_{r_{11}}}{dE} + \mid r_{21} \mid^2 \frac{d\theta_{r_{21}}}{dE} \right] \tag{4.3}$$

and

$$\rho^R(E, 1) = \frac{1}{2\pi}\left[\mid t_{11} \mid^2 \frac{d\theta_{t_{11}}}{dE} + \mid t_{21} \mid^2 \frac{d\theta_{t_{21}}}{dE} \right] \tag{4.4}$$

That is, $\rho^L(E, 1)$ consist of reflection channels and $\rho^R(E, 1)$ consist of transmission channels. The correction term corresponding to Eq. 3.59 depends on parameters of the incident channel only and gives the following identity.

$$\rho(E, 1) = \rho^L(E, 1) + \rho^R(E, 1) + \frac{1}{2\pi} \frac{m_0 \mid r_{11} \mid}{\hbar k_1^2} sin(\theta_{r_{11}}) \tag{4.5}$$

While this was known in 1D (Eq. 3.59) earlier, the explicit calculation in this chapter reveals that this is also true for the multichannel situation and the correction term depends solely on the input channel parameters, provided we account for the evanescent modes to calculate them. In Sects. 4.1 and 4.2 we will demonstrate how these evanescent modes are included. Similarly, semi-classical injectance for incidence along left lead in k_2 channel is

$$\rho(E, 2) \approx \frac{1}{2\pi}\left[\mid r_{22} \mid^2 \frac{d\theta_{r_{22}}}{dE} + \mid r_{12} \mid^2 \frac{d\theta_{r_{12}}}{dE} + \mid t_{22} \mid^2 \frac{d\theta_{t_{22}}}{dE} + \mid t_{12} \mid^2 \frac{d\theta_{t_{12}}}{dE} \right] \tag{4.6}$$

We can break this up as

$$\rho(E, 2) \approx \rho^L(E, 2) + \rho^R(E, 2) \tag{4.7}$$

where

$$\rho^L(E, 2) = \frac{1}{2\pi}\left[\mid r_{22} \mid^2 \frac{d\theta_{r_{22}}}{dE} + \mid r_{12} \mid^2 \frac{d\theta_{r_{12}}}{dE} \right] \tag{4.8}$$

and

$$\rho^R(E, 2) = \frac{1}{2\pi}\left[\mid t_{22} \mid^2 \frac{d\theta_{t_{22}}}{dE} + \mid t_{12} \mid^2 \frac{d\theta_{t_{12}}}{dE} \right] \tag{4.9}$$

and with the correction term we get

$$\rho(E, 2) = \rho^L(E, 2) + \rho^R(E, 2) + \frac{1}{2\pi} \frac{m_0 \mid r_{22} \mid}{\hbar k_2^2} sin(\theta_{r_{22}}) \tag{4.10}$$

The RHS of Eqs. 4.1 and 4.6 can be determined from the S-matrix. This S-matrix approach is easily accessible to experimentalists. But the correction term can be very large in mesoscopic systems that are in the quantum regime. The physical interpretation of the correction term as discussed after Eq. 3.59 also needs a major change.

4.1 A Typical Scattering Problem in Q1D

An analytic treatment of scattering in Q1D for a general system of the type shown in Fig. 2.8 is given in this section in terms of partial wave analysis. The purpose is to explicitly demonstrate the spectrum schematically shown in Fig. 4.2d, leading to general occurrence of Fano resonances which will be accompanied by transmission zeroes and associated discontinuous π phase shifts. The reader will further understand in the next few chapters that this will lead to negative partial density of states and time travel. The analysis of this section can be adopted to solve the system in Fig. 4.1 explicitly and has been done earlier [5]. Consider the scattering potential in the shaded region of Fig. 4.1 to be $U_g(x, y)$. Left reservoir acts as source of electrons and right reservoir as sink (not shown in Fig. 4.1). As usual the x-direction is along the source to sink direction and the y-direction is perpendicular to it. The confinement potential in the leads, i.e., in regions I and II, is taken to be hard wall in the y-direction (or, transverse direction), given by

$$\begin{aligned} U(x, y) = U(y) &= \infty \text{ for } \mid y \mid \geq \frac{b}{2} \text{ and } \mid x \mid \geq a \\ &= 0 \text{ for } \mid y \mid < \frac{b}{2} \text{ and } \mid x \mid \geq a \end{aligned} \tag{4.11}$$

The potential in the shaded region is of the form

$$\begin{aligned} U(x, y) &= \infty \text{ for } \mid y \mid \geq \frac{c}{2} \text{ and } \mid x \mid < a \\ &= U_g(x, y) \text{ for } \mid y \mid < \frac{c}{2} \text{ and } \mid x \mid < a \end{aligned} \tag{4.12}$$

Thus we have taken the system to be embedded in 2D and the reason for this is that the third direction is determined by the way the 2DEG (two-dimensional electron gas) is made and usually there is only one quantized energy mode for this degree of freedom, which can be set to zero by the choice of reference frame of energy.

We will first solve the Schrodinger equation in the leads and then match them to the solutions in the sample region which is the shaded area in Fig. 4.1. The Schrödinger equation in the two-dimensional lead is

$$\frac{-\hbar^2}{2m_0}\frac{\partial^2}{\partial x^2}\psi(x, y) - \frac{\hbar^2}{2m_0}\frac{\partial^2}{\partial y^2}\psi(x, y) + U(y)\psi(x, y) = E\psi(x, y) \tag{4.13}$$

Since the potential depends only on y, we can separate as follows.

$$\psi(x, y) = c_m(x)\chi_m(y) \tag{4.14}$$

From Eq. 4.13 and 4.14, we get $$\left[\frac{-\hbar^2}{2m_0}\frac{d^2}{dy^2} + U(y)\right]\chi_m(y) = \epsilon_m \chi_m(y) \quad (4.15)$$

and $$\frac{-\hbar^2}{2m_0}\frac{d^2}{dx^2}c_m(x) = (E - \epsilon_m)c_m(x) \quad (4.16)$$

From Eq. 4.15 $$\chi_m(y) = \sqrt{\frac{2}{b}}Sin\left[\frac{m\pi}{b}(y + \frac{b}{2})\right] \qquad \epsilon_m = \frac{\hbar^2 m^2 \pi^2}{2m_0 b^2} \quad (4.17)$$

For $m = 1$ in Eq. 4.17, we get the dotted curve in the leads of Fig. 2.8. For $m = 2$ in Eq. 4.17, we get the solid curve in the leads of Fig. 2.8. For $m = 3$, we get the dashed curve in the leads of Fig. 2.8. And so on.

From Eq. 4.16 $$c_m(x) = e^{\pm ik_m x} \qquad (E - \epsilon_m) = \frac{\hbar^2 k_m^2}{2m_0} \quad (4.18)$$

For $m = 1$ in Eq. 4.18, we get the dotted parabolic curve in Fig. 4.2d which is a plot of E versus k_1. For $m = 2$ in Eq. 4.18, we get the dashed parabolic curve in Fig. 4.2d which is a plot of E versus k_2. For $m = 3$ we get the solid parabolic curve in Fig. 4.2d. We have not yet obtained the bound states shown by the dotted straight lines or dashed straight lines, or solid straight lines in Fig. 4.2d. For them we have to solve for the Schrodinger equation in the shaded region of Fig. 4.1 which will be done from Eq. 4.30 onward.

From Eqs. 4.17 and 4.18 $$E = \frac{\hbar^2 m^2 \pi^2}{2m_0 b^2} + \frac{\hbar^2 k_m^2}{2m_0} = E_{m,k_m} \text{ (say)} \quad (4.19)$$

Consistent with Eqs. 4.14, 4.17, and 4.18 we can write a general solution as a linear combination of $\frac{1}{\sqrt{k_m}}\chi_m(y)e^{\pm ik_m x}$, where $s_{nm}^{e/o}$ are coefficients to be determined. Therefore,

$$\psi(x, y) = \quad (4.20)$$

$$= \frac{1}{2\sqrt{k_1}}\left[\delta_{11}e^{-ik_1 x} - s_{11}^e e^{ik_1 x}\right]\chi_1(y) + \frac{1}{2\sqrt{k_2}}\left[\delta_{21}e^{-ik_2 x} - s_{21}^e e^{ik_2 x}\right]\chi_2(y) + ...$$

$$+ \frac{1}{2\sqrt{k_1}}\left[\delta_{11}e^{-ik_1 x} - s_{11}^o e^{ik_1 x}\right]\chi_1(y) + \frac{1}{2\sqrt{k_2}}\left[\delta_{21}e^{-ik_2 x} - s_{21}^o e^{ik_2 x}\right]\chi_2(y) + ...$$

Each term in Eq. 4.20 satisfies Eq. 4.13 and so their sum also satisfies Eq. 4.13 because it is a linear differential equation. For example $\frac{1}{2\sqrt{k_1}}s_{11}^e e^{ik_1 x}\chi_1(y)$ satisfy Eq. 4.13 with $E = \frac{\hbar^2 1^2 \pi^2}{2m_0 b^2} + \frac{\hbar^2 k_1^2}{2m_0}$ while $\frac{1}{2\sqrt{k_2}}s_{21}^o e^{ik_2 x}\chi_2(y)$ satisfy Eq. 4.13 with $E = \frac{\hbar^2 2^2 \pi^2}{2m_0 b^2} + \frac{\hbar^2 k_2^2}{2m_0}$. The reader can find some more terms or break up the terms above in to some strange terms and mathematically there is nothing wrong. However, Eq. 4.20 is supposed to correspond to some physical situation and we have to address that. Equation 4.20 is essentially a series

$$\psi(x, y) = \sum_{m=1}^{\infty} \frac{1}{2\sqrt{k_m}} \left[\delta_{mm'} e^{-ik_m x} - s^e_{mm'} e^{ik_m x} \right] \chi_m(y)$$

$$+ \sum_{m=1}^{\infty} \frac{1}{2\sqrt{k_m}} \left[\delta_{mm'} e^{-ik_m x} - s^o_{mm'} e^{ik_m x} \right] \chi_m(y) \tag{4.21}$$

$$= \frac{\psi^e(x, y) + \psi^o(x, y)}{2} \tag{4.22}$$

where $\psi^e(x, y)$ is an even function of x and $\psi^o(x, y)$ is an odd function of x. This can always be demanded by adjusting the unknown coefficients $s^e_{mm'}$ and $s^o_{mm'}$ because any arbitrary function can be decomposed into a sum of even and odd function, and any matrix into a sum of symmetric and anti-symmetric matrix (see Eq. 4.24). Special situation here is that because the Schrodinger equation is a linear differential equation, both $\psi^e(x, y)$ and $\psi^o(x, y)$ independently satisfy Eq. 4.13. In Eq. 4.20 we can shuffle around the terms in the series and add them because the Hilbert space is complete by the axioms. This implies that each term in the series is in Hilbert space and the entire series is also in Hilbert space and so different terms can be added. Note that these additions have the spirit of vector addition as the Hilbert space is a normed vector space.

$\psi(x, y)$ of Eq. 4.21 is valid for $|x| \geq a$ and hence also valid for $x < -a$. Therefore, to find the wave-function in the left lead, we just have to use the coordinate value $-x$ in Eq. 4.21. For $x < -a$, $\psi(x, y)$ should correspond to a physical situation in the left lead of Fig. 4.1.

$$\psi(x, y) = \sum_{m=1}^{\infty} \frac{1}{2\sqrt{k_m}} \left[\delta_{mm'} e^{ik_m x} - s^o_{mm'} e^{-ik_m x} \right] \chi_m(y)$$

$$+ \sum_{m=1}^{\infty} \frac{1}{2\sqrt{k_m}} \left[\delta_{mm'} e^{ik_m x} - s^e_{mm'} e^{-ik_m x} \right] \chi_m(y) \tag{4.23}$$

$$= \frac{1}{\sqrt{k_{m'}}} e^{ik_{m'} x} \chi_{m'}(y) + \sum_{m=1}^{\infty} \frac{1}{\sqrt{k_m}} \frac{(-s^o_{mm'} - s^e_{mm'})}{2} e^{-ik_m x} \chi_m(y) \tag{4.24}$$

Note that Eq. 4.24 is only valid for $x < -a$ and to obtain its value at some $x < -a$ we have to only use the value $x = |x|$ because the negative sign has been already accounted for in going from Eq. 4.21 to Eq. 4.23. Indeed when we match the boundary conditions in Eqs. 4.35 and 4.45, this is what we will do, as is evident from the discussions in the paragraph after Eq. 4.25 where we explicitly keep track of the sign of x, leading to Eq. 4.34. Let us compare with the usual way in which we write the wave-function in 1D as $e^{ikx} + re^{-ikx}$. Like in 1D, we have in mind the following physical situation. That an ensemble of quantum particles is incident from the left. This incident ensemble in the left lead can be described by a wave-packet whose components are $e^{ik_{m'}x}\chi_{m'}(y)$ and there is a reflected wave-packet whose components are $\tilde{r}_{mm'}e^{-ik_m x}\chi_m(y)$. That $|\tilde{r}_{mm'}|^2$ can be read off from Eq. 4.24 and stated below in Eq. 4.25 can be interpreted as a fraction of the incident probability in the incoming m'th channel that is reflected to the mth channel. In Eq. 4.24 the incident wave-packet has Fourier components $\frac{1}{\sqrt{k_{m'}}}e^{ik_{m'}x}\chi_{m'}(y)$ and several reflected wave-packets in different channels indexed by m whose Fourier components are $\frac{1}{\sqrt{k_m}}\frac{(-s^o_{mm'}-s^e_{mm'})}{2}e^{-ik_m x}\chi_m(y)$.

$$\tilde{r}_{mm'} = -\frac{(s^o_{mm'} + s^e_{mm'})}{2} \tag{4.25}$$

Therefore, $\tilde{r}_{mm'}$ is the reflection amplitude from the m'th incident channel to the mth channel in the left lead as Eq. 4.24 is defined only for $x < -a$.

Now let us come to the right lead, that is $x > a$. Once again we can start from Eq. 4.21 that is valid for $|x| \geq a$. To find the wave-function at $x = +a$ we have to just substitute the coordinate value $x = +a$ in Eq. 4.21. But we can use symmetry to do some simplification. We can try to find out the value of $\psi(x, y)$ at $x = a$ from its value at $x = -a$. Let us substitute $x \rightarrow -x$ in Eq. 4.21 and note that this is how we can relate the wave-functions at positive and negative x. That means to find the wave-function at $x = +a$ we first find its simplified value at $x = -a$, because the symmetry allows us to do some simplification. However, this simplification scheme can only be applied at one end (that is the left end) because the whole purpose is to get the boundary conditions at the two ends from one end. So there are two alternate approaches to this scattering problem. In this approach we apply or calculate boundary conditions at only one end twice, once without applying symmetry and then by applying symmetry. In the other approach we do not use any symmetry but apply or calculate boundary condition at both ends separately. Because the wave-function at the two ends is always different and something has to be done differently at the two ends. Therefore, noting that the second series in Eq. 4.21 is odd, we get

$$\begin{aligned}\psi(x, y) = &\sum_{m=1}^{\infty} \frac{1}{2\sqrt{k_m}}\left[\delta_{mm'}e^{ik_m x} - s^e_{mm'}e^{-ik_m x}\right]\chi_m(y) \\ &-\sum_{m=1}^{\infty} \frac{1}{2\sqrt{k_m}}\left[\delta_{mm'}e^{ik_m x} - s^o_{mm'}e^{-ik_m x}\right]\chi_m(y)\end{aligned} \tag{4.26}$$

$$= \sum_{m=1}^{\infty} \frac{1}{\sqrt{k_m}}\frac{(-s^e_{mm'} + s^o_{mm'})}{2}e^{-ik_m x}\chi_m(y) \tag{4.27}$$

In this simplified value we re-substitute $x \to -x$ and in the expression that will be obtained, x can only admit positive values of $x > a$. So

$$\psi(x, y) = \sum_{m=1}^{\infty} \frac{1}{\sqrt{k_m}} \frac{(-s^e_{mm'} + s^o_{mm'})}{2} e^{ik_m x} \chi_m(y) \tag{4.28}$$

$$\tilde{t}_{mm'} = \frac{(s^o_{mm'} - s^e_{mm'})}{2} \tag{4.29}$$

Again in comparison to a 1D scattering problem, $\tilde{t}_{mm'}$ is the transmission amplitude for an electron incident from left along m' channel and transmitted to the mth channel on the right.

So far we were solving the Schrodinger equation in the leads. Now let us come to the scattering region, that is the shaded region in Fig. 4.1. Wave-function in the shaded region, that is in the potential region where $U(x, y) = U_g(x, y)$, can be written as

$$\eta^e(x, y) = \sum_{n=1}^{\infty} c_n \zeta^e_n(x, y) \tag{4.30}$$

$$\eta^o(x, y) = \sum_{n=1}^{\infty} c_n \zeta^o_n(x, y) \tag{4.31}$$

as internal wave-function too can be written as

$$\eta(x, y) = \frac{\eta^e(x, y) + \eta^o(x, y)}{2} \tag{4.32}$$

From Eqs. 4.21 and 4.22 (valid for all $|x| \geq a$) we get

$$\psi^e(x, y) = \sum_{m=1}^{\infty} \frac{1}{\sqrt{k_m}} \left[\delta_{mm'} e^{-ik_m x} - s^e_{mm'} e^{ik_m x} \right] \chi_m(y) \tag{4.33}$$

It is also natural to introduce an index m' on the LHS to make it compatible with the RHS.

$$\psi^e_{m'}(x, y) = \sum_{m=1}^{\infty} \frac{1}{\sqrt{k_m}} \left[\delta_{mm'} e^{-ik_m x} - s^e_{mm'} e^{ik_m x} \right] \chi_m(y) \tag{4.34}$$

Thus Eq. 4.34 is valid for $x = +a$ (see Fig. 4.1). Now we can equate ψ^e of Eq. 4.34 with η^e of Eq. 4.30 and similarly we can equate ψ^o with η^o, which will match ψ of Eq. 4.22 to η of Eq. 4.32 (crucially depends on axiomatic completeness of Hilbert space). But first let us apply current conserving boundary condition to get

$$\frac{\partial \psi^e_{m'}(x,y)}{\partial x}|_{x=a} = \frac{\partial \eta^e(x,y)}{\partial x}|_{x=a} \tag{4.35}$$

There are two quantum numbers involved that has to be respected. One is e that comes from spatial symmetry in x-direction and the other is m' coming from sub-band quantization. So on the RHS of Eq. 4.35 too we introduce the index m'. Note that if the system did not have the symmetry of $x \to -x$ then we cannot match ψ^e to η^e alone but we will have to match ψ^e to the full linear combination given by η in Eq. 4.32.

$$\frac{\partial \psi^e_{m'}(x,y)}{\partial x}|_{x=a} = \frac{\partial \eta^e_{m'}(x,y)}{\partial x}|_{x=a} \tag{4.36}$$

Substituting from Eqs. 4.30 and 4.34 into Eq. 4.35

$$\sum_{m=1}^{\infty} \sqrt{\frac{1}{k_m}} \left(\delta_{mm'}(-ik_m)e^{-ik_m a} - s^e_{mm'}(ik_m)e^{ik_m a} \right) \chi_m(y) = \sum_{n=1}^{\infty} c_n \frac{\partial \zeta^e_n(a,y)}{\partial x} \tag{4.37}$$

One has to first do the derivative and then substitute $x = a$. Using $\chi_m(y)$ from Eq. 4.17 we get

$$-i \sum_{m=1}^{\infty} \sqrt{\frac{2k_m}{b}} \left(\delta_{mm'}e^{-ik_m a} + s^e_{mm'}e^{ik_m a} \right) Sin\frac{m\pi}{b}\left(y + \frac{b}{2}\right) = \sum_{n=1}^{\infty} c_n \frac{\partial \zeta^e_n(a,y)}{\partial x} \tag{4.38}$$

Multiplying by $(\sqrt{\frac{2}{b}}) Sin\frac{m''\pi}{b}(y + \frac{b}{2})$ from the left on both sides, integrating and using

$$\int_{-\frac{b}{2}}^{\frac{b}{2}} Sin\frac{m''\pi}{b}\left(y + \frac{b}{2}\right) Sin\frac{m\pi}{b}\left(y + \frac{b}{2}\right) dy = \frac{b}{2}\delta_{m''m} \tag{4.39}$$

we get

$$\begin{aligned} &-i \sum_{m=1}^{\infty} \sqrt{k_m} \left(\delta_{mm'}e^{-ik_m a} + s^e_{mm'}e^{ik_m a} \right) \delta_{m''m} \\ &= \sum_{n=1}^{\infty} c_n \sqrt{\frac{2}{b}} \int_{-\frac{b}{2}}^{\frac{b}{2}} Sin\frac{m''\pi}{b}\left(y + \frac{b}{2}\right) \frac{\partial \zeta^e_n(a,y)}{\partial x} dy \end{aligned} \tag{4.40}$$

$$\begin{aligned} \text{or } &-i\sqrt{k_{m''}} \left(\delta_{m''m'}e^{-ik_{m''} a} + s^e_{m''m'}e^{ik_{m''} a} \right) \\ &= \sum_{n=1}^{\infty} c_n \sqrt{\frac{2}{b}} \int_{-\frac{b}{2}}^{\frac{b}{2}} Sin\frac{m''\pi}{b}\left(y + \frac{b}{2}\right) \frac{\partial \zeta^e_n(a,y)}{\partial x} dy \end{aligned} \tag{4.41}$$

$$\text{Let } \sqrt{\frac{2}{b}} \int_{-\frac{b}{2}}^{\frac{b}{2}} Sin\frac{m''\pi}{b}(y+\frac{b}{2})\frac{\partial \zeta_n^e(a,y)}{\partial x}dy = F_{m''n}^e\sqrt{k_{m''}k_n} \tag{4.42}$$

Therefore,

$$-i\sqrt{k_{m''}}\left(\delta_{m''m'}e^{-ik_{m''}a} + s_{m''m'}^e e^{ik_{m''}a}\right) = \sum_{n=1}^{\infty} c_n F_{m''n}^e \sqrt{k_{m''}k_n} \tag{4.43}$$

Now we interchange the indices of the delta function only and change the dummy index on RHS as $n \to m$ as there is no m on LHS.

$$-i\sqrt{k_{m''}}\left(\delta_{m'm''}e^{-ik_{m''}a} + s_{m''m'}^e e^{ik_{m''}a}\right) = \sum_{m=1}^{\infty} c_m F_{m''m}^e \sqrt{k_{m''}k_m} \tag{4.44}$$

Now we apply boundary condition to match wave-functions at $x = a$.

$$\psi_{m'}^e(x,y)|_{x=a} = \eta_{m'}^e(x,y)|_{x=a} \tag{4.45}$$

Substituting from Eqs. 4.30, 4.34, 4.17 into Eq. 4.45

$$\sum_{m=1}^{\infty}\sqrt{\frac{2}{bk_m}}\left(\delta_{mm'}e^{-ik_m a} - s_{mm'}^e e^{ik_m a}\right)Sin\frac{m\pi}{b}(y+\frac{b}{2}) = \sum_{n=1}^{\infty} c_n\zeta_n^e(a,y) \tag{4.46}$$

We multiply $(\sqrt{\frac{2}{b}})Sin\frac{m''\pi}{b}(y+\frac{b}{2})$ on both sides from the left and substitute

$$\zeta_n^e(a,y) = \sqrt{\frac{2}{b}}Sin\frac{n\pi}{b}\left(y+\frac{b}{2}\right) \tag{4.47}$$

because the internal and lead wave-functions have to match on the dotted line at $x = a$ in Fig. 4.1. After this substitution we integrate both sides and use Eq. 4.39 to get

$$\sum_{m=1}^{\infty}\frac{2}{b\sqrt{k_m}}\left(\delta_{mm'}e^{-ik_m a} - s_{mm'}^e e^{ik_m a}\right)\frac{b}{2}\delta_{m''m} = \sum_{n=1}^{\infty} c_n(\frac{2}{b})(\frac{b}{2})\delta_{m''n} \tag{4.48}$$

or

$$\sum_{m=1}^{\infty}\frac{1}{\sqrt{k_m}}e^{-ik_m a}\delta_{m''m}\delta_{mm'} - \sum_{m=1}^{\infty}\frac{1}{\sqrt{k_m}}\delta_{m''m}s_{mm'}^e e^{ik_m a} = \sum_{n=1}^{\infty} c_n\delta_{m''n} \tag{4.49}$$

$$\text{or } \frac{1}{\sqrt{k_{m''}}} e^{-ik_{m''}a} \delta_{m''m'} - \sum_{m=1}^{\infty} \frac{1}{\sqrt{k_m}} \delta_{m''m} s^e_{mm'} e^{ik_m a} = \sum_{n=1}^{\infty} c_n \delta_{m''n} \tag{4.50}$$

$$\text{or } \frac{1}{\sqrt{k_{m''}}} \left(\delta_{m''m'} e^{-ik_{m''}a} - s^e_{m''m'} e^{ik_{m''}a} \right) = c_{m''} \tag{4.51}$$

Interchanging m'' and m' in the delta function only, in Eq. 4.51

$$\frac{1}{\sqrt{k_{m''}}} \left(\delta_{m'm''} e^{-ik_{m''}a} - s^e_{m''m'} e^{ik_{m''}a} \right) = c_{m''} \tag{4.52}$$

From Eq. 4.52 with $m'' \to m$

$$\frac{1}{\sqrt{k_m}} \left(\delta_{m'm} e^{-ik_m a} - s^e_{mm'} e^{ik_m a} \right) = c_m \tag{4.53}$$

Substituting this c_m in Eq. 4.44.

$$\begin{aligned} &- i\sqrt{k_{m''}} \left(\delta_{m'm''} e^{-ik_{m''}a} + s^e_{m''m'} e^{ik_{m''}a} \right) \\ = &\sum_{m=1}^{\infty} \frac{1}{\sqrt{k_m}} \left(\delta_{m'm} e^{-ik_m a} - s^e_{mm'} e^{ik_m a} \right) F^e_{m''m} \sqrt{k_{m''} k_m} \end{aligned} \tag{4.54}$$

$$\text{or } \sum_{m=1}^{\infty} (F^e_{m''m} - i\delta_{m''m}) e^{ik_m a} s^e_{mm'} = F^e_{m''m'} e^{-ik_{m'}a} + i\delta_{m'm''} e^{-ik_{m''}a} \tag{4.55}$$

Note that the last term on the RHS is a diagonal term (with respect to symmetry) because of the Kronecker delta, there being no sum. So in this last term alone we can interchange m'' and m' and it still remains a diagonal term.

$$\sum_{m=1}^{\infty} (F^e_{m''m} - i\delta_{m''m}) e^{ik_m a} s^e_{mm'} = F^e_{m''m'} e^{-ik_{m'}a} + i\delta_{m''m'} e^{-ik_{m'}a} \tag{4.56}$$

$$\text{or } \sum_{m=1}^{\infty} \left(F^e_{m''m} - i\delta_{m''m} \right) e^{ik_m a} s^e_{mm'} = \left(F^e_{m''m'} + i\delta_{m''m'} \right) e^{-ik_{m'}a} \tag{4.57}$$

This is the equation that we have to solve to find $s^e_{mm'}$. Similarly we will get an equation for $s^o_{mm'}$. Multiplying identical terms on both sides of Eq. 4.57 we get

$$\sum_{n=1}\left[(F^e - iI)^{-1}\right]_{nm''}\sum_{m=1}^{\infty}(F^e_{m''m} - i\delta_{m''m})e^{ik_m a}s^e_{mm'}$$
$$= \sum_{n=1}\left[(F^e - iI)^{-1}\right]_{nm''}(F^e_{m''m'} + i\delta_{m''m'})e^{-ik_{m'}a} \tag{4.58}$$

$$\text{or } \sum_{nm}\left[(F^e - iI)^{-1}\right]_{nm''}(F^e_{m''m} - i\delta_{m''m})e^{ik_m a}s^e_{mm'}$$
$$= \sum_{n}\left[(F^e - iI)^{-1}\right]_{nm''}\left[F^e + iI\right]_{m''m'}e^{-ik_{m'}a} \tag{4.59}$$

$$\text{or } \sum_{nm}\left[(F^e - iI)^{-1}\right]_{nm''}\left[F^e - iI\right]_{m''m}e^{ik_m a}s^e_{mm'}$$
$$= \sum_{n}\left[(F^e - iI)^{-1}\right]_{nm''}\left[F^e + iI\right]_{m''m'}e^{-ik_{m'}a} \tag{4.60}$$

Introducing a sum on both sides

$$\text{or } \sum_{nm}\sum_{m''}\left[(F^e - iI)^{-1}\right]_{nm''}\left[F^e - iI\right]_{m''m}e^{ik_m a}s^e_{mm'}$$
$$= \sum_{n}\sum_{m''}\left[(F^e - iI)^{-1}\right]_{nm''}\left[F^e + iI\right]_{m''m'}e^{-ik_{m'}a} \tag{4.61}$$

In quantum mechanics, a series can be seen as a limit of a finite sum and so we can talk of Eq. 4.57 as a matrix equation. In this particular case the series has a natural truncation because very high k_m values correspond to very high energy states that are not possible in condensed matter systems. The work function of a material provides a natural cut off. Since on LHS, after the sum over m'', we are multiplying matrix elements from two matrices that are inverse to each other, two matrix elements must multiply to give a 0 or a 1.

$$\sum_{nm}\delta_{nm}e^{ik_m a}s^e_{mm'} = \sum_{n}\sum_{m''}\left[(F^e - iI)^{-1}\right]_{nm''}\left[F^e + iI\right]_{m''m'}e^{-ik_{m'}a} \tag{4.62}$$

$$\text{or } \sum_{n}e^{ik_n a}s^e_{nm'} = \sum_{n}\sum_{m''}\left[(F^e - iI)^{-1}\right]_{nm''}\left[F^e + iI\right]_{m''m'}e^{-ik_{m'}a} \tag{4.63}$$

$$\text{or} \sum_n e^{ik_n a} s^e_{nm'} = \sum_n \Big[(F^e - iI)^{-1}(F^e + iI)\Big]_{nm'} e^{-ik_{m'}a} \tag{4.64}$$

Since this is true for any arbitrary $k_n a$, we can remove the sum from both sides.

$$e^{ik_n a} s^e_{nm'} = \Big[(F^e - iI)^{-1}(F^e + iI)\Big]_{nm'} e^{-ik_{m'}a} \tag{4.65}$$

$$\text{or } s^e_{nm'} = e^{-ik_n a}\Big[(F^e - iI)^{-1}(F^e + iI)\Big]_{nm'} e^{-ik_{m'}a} \tag{4.66}$$

Now binomial expansion like procedure can be followed because the identity matrix commutes with any matrix.

$$\begin{aligned}(F^e - iI)^{-1}(F^e + iI) &= \frac{1}{(F^e - iI)}(F^e + iI) = \frac{1}{i(-iF^e - I)} i(-iF^e + I)\\ &= \frac{1}{(-iF^e - I)}(-iF^e + I) = \frac{-1}{(I + iF^e)}(I - iF^e)\end{aligned} \tag{4.67}$$

$$= -[1 - iF^e + (iF^e)^2 - (iF^e)^3 + \ldots\ldots\ldots\ldots](1 - iF^e) \tag{4.68}$$

$$= [-1 + iF^e - (iF^e)^2 + (iF^e)^3 \ldots\ldots\ldots + iF^e - (iF^e)^2 + (iF^e)^3 - (iF^e)^4 + \ldots.] \tag{4.69}$$

$$= \Big[-1 + 2(iF^e) - 2(iF^e)^2 + 2(iF^e)^3 - 2(iF^e)^4 + \ldots.\Big] \tag{4.70}$$

$$= \Big[1 - 2\Big(1 - (iF^e) + (iF^e)^2 - (iF^e)^3 + (iF^e)^4 + \ldots.\Big)\Big] \tag{4.71}$$

$$= \Big[1 - 2(1 + iF^e)^{-1}\Big] \tag{4.72}$$

$$= 1 - \frac{2}{(1 + iF^e)} = 1 - \frac{2}{i}\frac{1}{(-i + F^e)} \tag{4.73}$$

$$= 1 - \frac{2}{i}(F^e - i)^{-1} = 1 + 2i(F^e - i)^{-1} \tag{4.74}$$

Thus from Eqs. 4.67 and 4.74

$$(F^e - iI)^{-1}(F^e + iI) = 1 + 2i(F^e - i)^{-1} \tag{4.75}$$

Therefore from Eq. 4.66, where we drop the superscript e for convenience, which we will bring back after some simplification.

$$s_{nm'} = e^{-ik_n a}\left[1 + 2i(F - iI)^{-1}\right]_{nm'} e^{-ik_{m'}a} \tag{4.76}$$

To explain the straight lines in Fig. 4.2d, representing bound states and resonant states, let us reconsider Eq. 4.19

$$E = \frac{\hbar^2 m^2 \pi^2}{2m_0 b^2} + \frac{\hbar^2 k_m^2}{2m_0} \tag{4.77}$$

If we take $m = 1$ and plot E versus k_1 then we get the dotted parabola in Fig. 4.2d where the minimum of the parabola is $\epsilon_1 = \frac{\hbar^2 1^2 \pi^2}{2m_0 b^2}$ and referred to as the propagating threshold for the m=1 channel. If we take $m = 2$ and plot E versus k_2 then we get the dashed parabola in Fig. 4.2d where the minimum of the parabola is $\epsilon_2 = \frac{\hbar^2 2^2 \pi^2}{2m_0 b^2}$ and referred to as the propagating threshold for the m=2 channel. There can be thus an infinite number of these parabolas and Fig. 4.2d depicts only three of them.

In mesoscopic physics an experimentalist can fix the incident energy E and the range in which it can vary. So suppose

$$\frac{\hbar^2 \pi^2}{2m_0 b^2} < E < \frac{\hbar^2 4\pi^2}{2m_0 b^2} \tag{4.78}$$

That is

$$\epsilon_1 < E < \epsilon_2 \tag{4.79}$$

If we substitute an E value in this range in Eq. 4.77 then we can see that k_1 is real and k_2, k_3, ... are all imaginary. What it physically means with respect to Fig. 4.2d is the following. It means the incident energy E is varying between the minimum of the dotted parabola and the minimum of the dashed parabola. In this regime only k_1 is a propagating mode and the rest are evanescent modes. For example if k_2 is imaginary then we can substitute $k_2 \to i\kappa_2$ and substitute in Eq. 4.24 to see this. However, in this energy range there can be some discrete real values of k_2 for which one can find solutions to Eqs. 4.57 or 4.76. These will give the bound states like the dashed lines in Fig. 4.2d. Bound states can be determined from the singularities of the matrix equation, that is from Eqs. 4.57 or 4.76 we can say that solutions to the following equation will give the bound states.

$$det\left[F_{cc}^e - i1\right] = 0 \tag{4.80}$$

As stated above, Eq. 4.57 can be seen as a matrix equation, where the RHS terms are known, and the coefficients of $s^e_{mm'}$ on LHS are also known. We can drop the superscript e for convenience, that will be introduced later. Here "cc" means evanescent channel (or closed channel, see Eq. 4.81) for which both k_m and $k_{m'}$ in Eq. 4.57 are imaginary. For these values of k_m and $k_{m'}$, we can still find solutions to Eq. 4.57 with $s^e_{mm'} \neq 0$ for $m = m'$ only for some discrete values of k_m. Note that for $m = 2$ these solutions will be for $E < \epsilon_2 = \frac{\hbar^2 4\pi^2}{2m_0 b^2}$. An electron in such a state gets trapped in an eigen-state with no reflection or transmission. To explain F_{cc} further we can partition the matrix F into its open and closed channel parts as follows.

$$F = \begin{bmatrix} F_{oo} & F_{oc} \\ F_{co} & F_{cc} \end{bmatrix} \tag{4.81}$$

The subscript "o" refers to open and "c" refers to closed channels. Suppose for m' and m taking values from the set $\{1, 2\}$, F_{oo} will be a 2×2 matrix while F_{cc} will be an infinite dimensional matrix that can be truncated at some point based on physical arguments. The completeness of the Hilbert space ensures that the truncation is physical. F_{oo} is a matrix whose elements are of the form $F_{mm'}$ where m and m' are indices for open channels. Similarly F_{cc} is a matrix whose elements are of the form $F_{mm'}$ where m and m' are indices for closed channels. Thus F_{oc} is a matrix whose elements are of the form $F_{mm'}$ where m is an index for an open channel while m' is an index for a closed channel. Similarly F_{co} is a matrix whose elements are of the form $F_{mm'}$ where m is an index for a closed channel while m' is an index for an open channel.

Thus we may write on dropping the superscript e which will be reintroduced later

$$\begin{bmatrix} F - i1 \end{bmatrix}^{-1} = \begin{bmatrix} A & B \\ C & D \end{bmatrix} \tag{4.82}$$

Therefore from Eq. 4.81

$$\begin{bmatrix} F_{oo} - i1 & F_{oc} \\ F_{co} & F_{cc} - i1 \end{bmatrix} \begin{bmatrix} A & B \\ C & D \end{bmatrix} = \begin{bmatrix} 1 & 0 \\ 0 & 1 \end{bmatrix} \tag{4.83}$$

From Eq. 4.83

$$(F_{oo} - i1)A + F_{oc}C = 1 \tag{4.84}$$

$$(F_{oo} - i1)B + F_{oc}D = 0 \tag{4.85}$$

$$F_{co}A + (F_{cc} - i1)C = 0 \tag{4.86}$$

$$F_{co}B + (F_{cc} - i1)D = 1 \tag{4.87}$$

Therefore from Eq. 4.86

$$C = -[F_{cc} - iI]^{-1} F_{co} A \tag{4.88}$$

Putting the value of C from Eq. 4.88 into Eq. 4.84

$$\left[F_{oo} - iI\right]A - F_{oc}\left[F_{cc} - iI\right]^{-1} F_{co} A = 1 \tag{4.89}$$

$$\text{or} \quad \left(\left[F_{oo} - iI\right] - F_{oc}\left[F_{cc} - iI\right]^{-1} F_{co}\right)A = 1 \tag{4.90}$$

$$\text{or} \quad A = \left(F_{oo} - iI - F_{oc}(F_{cc} - iI)^{-1} F_{co}\right)^{-1} \tag{4.91}$$

From Eqs. 4.82 we see that

$$A = \left[(F - iI)^{-1}\right]_{oo} \tag{4.92}$$

In Eq. 4.76, if n and m' are open channels, then substituting A from Eq. 4.92 in Eq. 4.76

$$s_{nm'} = e^{-ik_n a}\left[1 + 2iA\right]_{nm'} e^{-ik_{m'} a} \tag{4.93}$$

Substituting from Eq. 4.91 into Eq. 4.93

$$s_{nm'} = e^{-ik_n a}\left[1 + 2i\left(F_{oo} - iI - F_{oc}(F_{cc} - iI)^{-1} F_{co}\right)^{-1}\right]_{nm'} e^{-ik_{m'} a} \tag{4.94}$$

$$\text{or} \quad s_{nm'} = e^{-ik_n a}\left[1 + 2i\left(G - iI\right)^{-1}\right]_{nm'} e^{-ik_{m'} a} \tag{4.95}$$

$$\text{where} \quad G = \left[F_{oo} - F_{oc}(F_{cc} - iI)^{-1} F_{co}\right] \tag{4.96}$$

From Eq. 4.95 and comparing with Eq. 4.75

$$s_{nm'} = e^{-ik_n a}\left[(G - iI)^{-1}(G + iI)\right]_{nm'} e^{-ik_{m'} a} \tag{4.97}$$

Suppose there is only one propagating channel in both the leads. Then we put $n = m' = 1$ in Eq. 4.97 and also from Eq. 4.96 we see that G is a number.

$$G = F_{11} - \sum_{m'=2,m=2} F_{1m}[(F_{cc} - i1)^{-1}]_{mm'} F_{m'1} \tag{4.98}$$

From Eq. 4.97

$$s_{11} = e^{-ik_1 a}\left[\frac{G + iI}{G - iI}\right] e^{-ik_1 a} \tag{4.99}$$

From Eqs. 4.95 and 4.98

$$s_{11} = e^{-2ik_1 a}\left[1 + 2i\left(F_{11} - \sum_{m',m=2} F_{1m}[(F_{cc} - i1)^{-1}]_{mm'} F_{m'1} - i1\right)^{-1}\right] \tag{4.100}$$

We reintroduce the superscript e to write

$$G^e = Cot\theta^e \text{ (say)} \tag{4.101}$$

From Eq. 4.99

$$s_{11}^e = e^{-2ik_1 a}\frac{Cot\theta^e + i1}{Cot\theta^e - i1} \tag{4.102}$$

$$\text{or } s_{11}^e = e^{-2ik_1 a}\frac{e^{i\theta^e}}{e^{-i\theta^e}} \tag{4.103}$$

$$\text{or } s_{11}^e = e^{-2ik_1 a} e^{2i\theta^e} = e^{2iArccot[G^e] - 2ik_1 a} \tag{4.104}$$

$$\text{or } s_{11}^e = e^{2i(\theta^e - k_1 a)} \tag{4.105}$$

$$\text{or } s_{11}^e = e^{2i\delta^e} \tag{4.106}$$

$$\text{where } \delta^e = \theta^e - k_1 a \tag{4.107}$$

$$\text{Similarly we can get } s_{11}^o = e^{2i\delta^o} \tag{4.108}$$

$$\text{and } \delta^o = \theta^o - k_1 a \tag{4.109}$$

According to the convention of Eq. 1.96 in Chap. 1, $\delta \equiv \theta_{t'}$, whereas we work with θ_t. Therefore

$$s_{11}^e = e^{2i\theta^e} \tag{4.110}$$

$$s_{11}^o = e^{2i\theta^o} \tag{4.111}$$

Now from Eqs. 4.29, we find that transmission amplitude ($\tilde{t}_{11}$) is given by

$$\tilde{t}_{11} = \frac{-s^e_{11} + s^o_{11}}{2} = \frac{-e^{2i\theta^e} + e^{2i\theta^o}}{2} \tag{4.112}$$

$$\text{or } \tilde{t}_{11} = \frac{1}{i} Sin(\theta^e - \theta^o)\left[Cos(\theta^e + \theta^o) + i Sin(\theta^e + \theta^o)\right]$$

We define new variables as

$$\phi = \theta^e - \theta^o \text{ and } \theta^r = \theta^e + \theta^o \tag{4.113}$$

$$\tilde{t}_{11} = -i Sin\phi e^{i\theta^r} \tag{4.114}$$

Similarly from Eq. 4.25 we find the reflection amplitude to be

$$\tilde{r}_{11} = -\frac{s^e_{11} + s^o_{11}}{2}$$

$$\text{or } \tilde{r}_{11} = -\frac{(e^{2i\theta^e} + e^{2i\theta^o})}{2}$$

$$\text{or } \tilde{r}_{11} = -Cos(\theta^e - \theta^o)\left[Cos(\theta^e + \theta^o) + i Sin(\theta^e + \theta^o)\right]$$

and

$$\tilde{r}_{11} = -Cos(\phi)e^{i\theta^r} \tag{4.115}$$

Thereby we get transmission amplitude and reflection amplitude as

$$\tilde{t}_{11} = -i Sin(\phi)e^{i\theta^r} \text{ and } \tilde{r}_{11} = -Cos(\phi)e^{i\theta^r} \tag{4.116}$$

One can see [6] for further details, but the discontinuous scattering phase shifts at the resonances or bound states were overlooked there. We can see them as follows. Since the system has reflection symmetry in x, bound states will be either an even state or an odd state. At an even bound state G^e will blow up as is clear from Eqs. 4.98 and 4.80. That is because the second term in G^e, as given by Eq. 4.98, has the inverse of a matrix and when the bound state condition given by Eq. 4.80 is satisfied, this second term will diverge. Therefore at a bound state $\theta^e = Arccot[G^e]$ (see Eq. 4.101) will go to zero. So at an even bound state $|\tilde{t}_{11}| = Sin(\phi) = Sin(\theta^e) = 0$ (see Eqs. 4.114 and 4.113) and we have seen in Chap. 2 that this leads to discontinuous phase shifts of π that will be further generalized in the next chapters.

The correction term of Eq. 4.5 in case of this system is $\frac{m_0|\tilde{r}_{11}|}{\hbar k_1^2}\sin(\theta^r) = \frac{m_0|\tilde{r}_{11}|}{\hbar k_1^2}\sin(\theta^e)$ (see Eq. 4.113), and therefore that too is zero from Eq. 4.101. Thus we have shown that the correction term in Eq. 4.5 is zero because $\sin(\theta^r) = 0$. At this point $| r |=| -Cos(\phi) |$ (see Eq. 4.115) $=| -Cos(\theta^e) |= Cos(0)$ (we have already shown $Sin(\theta_e) = 0$) and so $| r | \neq 0$. As stated earlier, if the correction term can be ignored or set to zero then the semi-classical formulas can be very useful to experimentalists.

4.2 Delta Function Potential in Q1D

We cannot solve the multichannel case (apart from truncating the series of evanescent modes and applying numerical techniques) for any potential except for the case described in this section. However, our main results can be proved generally using mathematical theorems and principles for any potential and this will be presented in the next two chapters. Here we will study the two propagating channels (with arbitrary number of evanescent channel) case for a particular potential, i.e., $U_g(x, y) = \gamma\delta(x)\delta(y - y_i)$, because it can be exactly solved and certain facts can be exhibited explicitly. In Fig. 4.3, the shaded region of Fig. 2.8 reduces to give a two-dimensional quantum wire with a delta function potential at position $(0, y_i)$ marked X. The dotted lines in Fig. 4.3 represent the fact that the quantum wire is connected to electron reservoirs.

We will first calculate this injectance from internal wave-function $\psi(x, y, 1)$ which allows one to make numerical verification of the semi-classical formulas in Eqs. 4.1 and 4.6 and their correction terms in Eqs. 4.5 and 4.10. We know that when incidence is along n=1 channel

$$\rho(1, E) = \int_{-\infty}^{\infty} dx \int_{-\frac{W}{2}}^{\frac{W}{2}} dy \sum_{k_1} | \psi(x, y, 1) |^2 \delta(E - E_{1,k_1}) \tag{4.117}$$

Here E_{1,k_1} is the energy of the incident electron along channel $n = 1$ with wave vector $k_{n=1}$ or k_1. Summing over n one can get the standard expression for DOS according to spectral theorem. Note that although k_1 can vary continuously, one can write sum over k_1 precisely because the electrons are countable. One can write for the system represented in Fig. 4.3 [3] when incidence is along n=1 channel

$$\psi(x, y, 1) = \sum_m f_m(x, 1)\chi_m(y) \tag{4.118}$$

Here $\chi_m(y)s$ are solutions in Eq. 4.17 in the leads in the transverse direction which is an infinite square well potential in y-direction. $\chi_m(y)s$ form a complete set and Eq. 4.118 is derived from the fact that at a given point x, the wave-function in the scattering region can be expanded in terms of $\chi_m(y)s$. We will present our results for the case of two propagating channels but the analysis and results are same for any number of channels. f_ms are generally of the form given below,

$$f_1(x,1) = e^{ik_1x} + r_{11}e^{-ik_1x} \quad for \quad x < 0 \tag{4.119}$$

$$= t_{11}e^{ik_1x} \quad for \quad x > 0 \tag{4.120}$$

$$f_2(x,1) = r_{21}e^{-ik_2x} \quad for \quad x < 0 \tag{4.121}$$

$$= t_{21}e^{ik_2x} \quad for \quad x > 0 \tag{4.122}$$

and for $m > 2$,

$$f_m(x,1) = r'_{m1}e^{\kappa_m x} \quad for \quad x < 0 \tag{4.123}$$

$$= t'_{m1}e^{-\kappa_m x} \quad for \quad x > 0 \tag{4.124}$$

For $m > 2$, $(\varepsilon_m - E) = \frac{\hbar^2\kappa_m^2}{2m_0}$, κ_m being positive define evanescent modes as now $E < \varepsilon_m$. r_{mn}, t_{mn} r'_{mn}, and t'_{mn} are unknowns to be determined. The scattering problem described above can be solved using mode matching technique. The reflection amplitudes are given by Bagwell [3],

$$r_{mn}(E) = -\frac{i\frac{\Gamma_{mn}}{2\sqrt{k_m k_n}}}{1 + \sum_e \frac{\Gamma_{ee}}{2\kappa_e} + i\sum_m \frac{\Gamma_{mm}}{2k_m}} \tag{4.125}$$

Γ_{mn} is the coupling strength between mth and nth modes, given by

$$\Gamma_{mn} = \gamma sin\left[\frac{m\pi}{W}\left(y_i + \frac{W}{2}\right)\right]sin\left[\frac{n\pi}{W}\left(y_i + \frac{W}{2}\right)\right] \tag{4.126}$$

The transmission amplitudes are given by

$$t_{mn} = r_{mn}(E) \;\; \text{for } m \neq n \tag{4.127}$$

$$\text{and } \; t_{mm}(E) = 1 + r_{mm}(E) \tag{4.128}$$

Along the same line one can show that

$$r'_{m1} = t'_{m1}(E) = -\frac{\frac{\Gamma_{m1}}{2\kappa_m}}{1 + \sum_e \frac{\Gamma_{ee}}{2\kappa_e} + i\left(\frac{\Gamma_{11}}{2k_1} + \frac{\Gamma_{22}}{2k_2}\right)} \tag{4.129}$$

$\sum_e$ denote sum over evanescent modes and run from m =3 to ∞, while $\sum_m$ denote the same for propagating modes (i.e., $m = 1$ and $m = 2$).

If the delta function potential is negative($\gamma < 0$), then there can be bound states as argued before. As explained in Fig. 4.2d it will always be a quasi-bound state. The quasi-bound state for channel $m = s \geq 2$ is given by Bagwell [3],

$$1+\sum_{e=s}^{\infty}\frac{\Gamma_{ee}}{2\kappa_e}=0 \tag{4.130}$$

Only bound state for $m = 1$ channel is a true bound state and it is given by the solution to the following equation, $1+\sum_{e=1}^{\infty}\frac{\Gamma_{ee}}{2\kappa_e}=0$. The bound state for $m = 2$ is given by

$$1+\sum_{e=2}^{\infty}\frac{\Gamma_{ee}}{2\kappa_e}=0 \tag{4.131}$$

At this energy we get a bound state for $m = 2$, but at that very energy $m = 1$ channel (or k_1 channel) is a propagating channel as can be seen in Eq. 4.78. Hence the bound state given by this Eq. 4.131 is a quasi-bound state or a resonance. The delta function in Eq. 4.117 summed over k_1 essentially yields a factor $\frac{1}{hv_1}$, where $v_1 = \frac{\hbar k_1}{m}$. Substituting for $\psi()$ from Eq. 4.118 and using the orthonormality of $\chi_m(y)s$, we get

$$\rho(1,E)=\frac{1}{hv_1}\Big[\int_{-\infty}^{\infty}dx\sum_m \mid f_m(x,1)\mid^2\Big] \tag{4.132}$$

$$=\frac{1}{hv_1}\Big[\int_{-\infty}^{\infty}dx\mid f_1(x,1)\mid^2+\int_{-\infty}^{\infty}dx\mid f_2(x,1)\mid^2+$$

$$\int_{-\infty}^{\infty}dx\mid f_3(x,1)\mid^2+\int_{-\infty}^{\infty}dx\mid f_4(x,1)\mid^2+....\Big] \tag{4.133}$$

Substituting the values of $f_m(x,1)s$ from Eqs. 4.119–4.124, we get

$$\rho(1,E)=\frac{1}{hv_1}\Big[\int_{-\infty}^{0}dx[1+\mid r_{11}\mid^2+2\mid r_{11}\mid Cos(2k_1x+\phi_1)]$$

$$+\int_{0}^{\infty}dx\mid t_{11}\mid^2+\int_{-\infty}^{0}dx\mid r_{21}\mid^2+\int_{0}^{\infty}dx\mid t_{21}\mid^2+H\Big] \tag{4.134}$$

Here, $r_{11}=\mid r_{11}\mid e^{-i\phi_1}$. Note that for $m > 2$,

$$H=\int_{-\infty}^{\infty}dx\sum_{m>2}\mid f_m(x,1)\mid^2=\sum_{m>2}\mid t'_{m1}\mid^2\Big[\int_{-\infty}^{0}e^{2\kappa_m x}dx+\int_{0}^{\infty}e^{-2\kappa_m x}dx\Big]$$

$$=\sum_{m>2}\frac{\mid t'_{m1}\mid^2}{\kappa_m} \tag{4.135}$$

Thus

$$\rho(E,1) = \frac{1+\mid r_{11}\mid^2+\mid r_{21}\mid^2}{hv_1}\int\limits_{-\infty}^{0} dx + \frac{\mid t_{11}\mid^2+\mid t_{21}\mid^2}{hv_1}\int\limits_{0}^{\infty} dx$$

$$+\frac{2\mid r_{11}\mid}{hv_1}\int\limits_{-\infty}^{0} dx Cos(2k_1x+\phi_1)$$

$$+\frac{1}{hv_1}\left(\frac{\mid t'_{31}\mid^2}{\kappa_3}+\frac{\mid t'_{41}\mid^2}{\kappa_4}+...\right) \tag{4.136}$$

Adding and subtracting the following terms,

$$\frac{\mid t_{11}\mid^2}{hv_1}\int\limits_{-\infty}^{0} dx, \frac{\mid t_{21}\mid^2}{hv_1}\int\limits_{-\infty}^{0} dx,$$

and using the fact that

$$\frac{\mid t_{11}\mid^2}{hv_1}\int\limits_{-\infty}^{0} dx = \frac{\mid t_{11}\mid^2}{hv_1}\int\limits_{0}^{\infty} dx$$

and $\mid r_{11}\mid^2+\mid r_{21}\mid^2+\mid t_{11}\mid^2+\mid t_{21}\mid^2 = 1$, we get

$$\rho(E,1) = \frac{1}{hv_1}\int\limits_{-\infty}^{\infty} dx + \frac{\mid r_{11}\mid}{hv_1}\int\limits_{-\infty}^{\infty} dx Cos(2k_1x+\phi_1)$$

$$+\frac{1}{hv_1}\left(\frac{\mid t'_{31}\mid^2}{\kappa_3}+\frac{\mid t'_{41}\mid^2}{\kappa_4}+......\right) \tag{4.137}$$

$\frac{1}{hv_1}\int\limits_{-\infty}^{\infty} dx = P_0(E)$ is the injectance in the absence of scatterer. Again if the scattering phase shift is defined with respect to the phase shift in absence of scatterer then this term can be dropped as shown in Eqs. 1.94 and 1.97. Now,

$$\int\limits_{-\infty}^{\infty} dx Cos(2k_1x+\phi_1) = \delta(2k_1)Cos(\phi_1)$$

As $k_1 = 0$ is not a propagating state contributing to transport, this term in Eq. 4.137 becomes zero, and we are left with

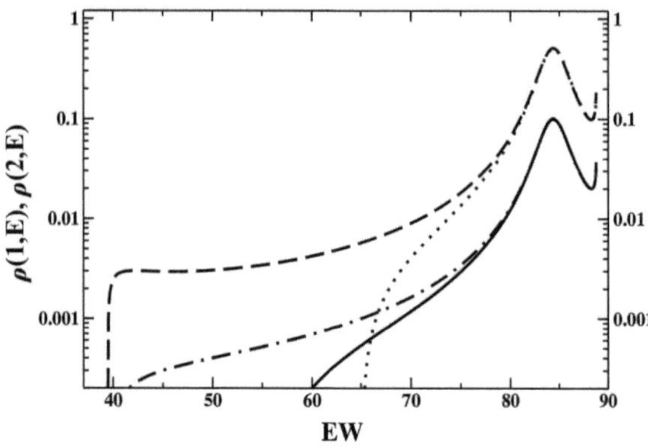

Fig. 4.4 The plot is of injectance versus incident energy EW. We have used $e\gamma W = -15$, and $y_i = 0.45W$. The figure shows that semi-classical formula becomes exact at resonance where there is a peak in injectance. Injectance of first transverse mode is less than that of the second

$$\rho(E, 1) = \frac{1}{hv_1}\left(\frac{|t'_{31}|^2}{\kappa_3} + \frac{|t'_{41}|^2}{\kappa_4} + ...\right) \tag{4.138}$$

Similarly, for incidence along channel 2, one can obtain the injectance.

$$\rho(E, 2) = \frac{1}{hv_2}\left(\frac{|t'_{32}|^2}{\kappa_3} + \frac{|t'_{42}|^2}{\kappa_4} + ...\right) \tag{4.139}$$

In Fig. 4.4 we make a comparison between the semi-classical injectance $\rho(E, 1)$ (solid curve) from Eq. 4.1 and the same from the exact formula in Eq. 4.138 (dash-dot curve). The resonance condition in Eq. 4.131 is satisfied at $EW = 84.29$ at which the injectance peaks. Surprisingly, at the resonance the semi-classical formula becomes exact. The same happens for $\rho(2, E)$ for which the semi-classical formula in Eq. 4.6 gives the dotted curve and the exact formula in Eq. 4.139 gives the dashed curve. That this is not a special situation for the delta function potential but a general feature of Fano resonances will be shown in the next chapter. We have also stated earlier that the exactness of a semi-classical formula can find applications which will be further elaborated in the next two chapters.

It is to be noted that the semi-classical formula becomes exact (not a good approximation but exact) in a quantum regime which will be proved in the next two sections and physical reasons will be given. But here we demonstrate in Fig. 4.5, plotted for the same parameters as in Fig. 4.4, the fact that at the resonance $EW = 84.29$ the phase fluctuations maximize. However, at the resonance the correction term goes to zero as $\theta_{r_{11}}$ goes to zero resulting in $Sin(\theta_{r_{11}})$ becoming zero in Eq. 4.5. The same happens for $\theta_{r_{22}}$. Hence this situation is different from what was thought earlier that only in the semi-classical regime when $|r_{11}|^2$ becomes negligible, the correction term can be ignored. In a semi-classical regime, particles are expected to behave like a classical particle which is entirely transmitted.

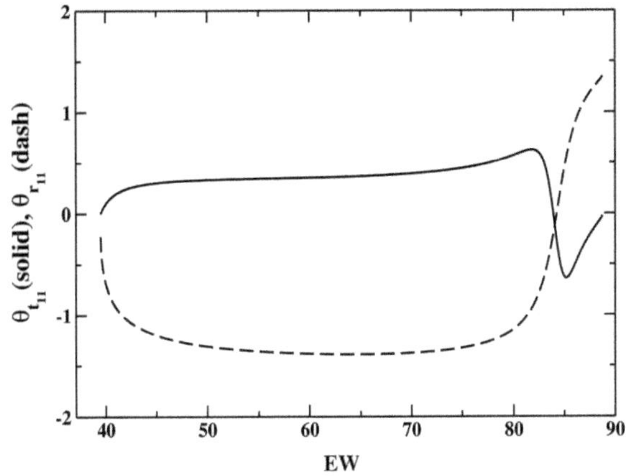

Fig. 4.5 Here we plot reflection phase shift (dashed curve) and transmission phase shift (solid curve) versus incident energy for the same parameter values as in Fig. 4.4. The figure shows that the scattering phase shifts become simultaneously zero at the resonance while being out of phase for the rest of the energy regime. And so the initial phase of the wave-function which is usually undetermined does not play a role. That initial phase can be absorbed in $\theta_{r_{11}}$ to give the same conclusion

References

1. Singha Deo P (2007) Phys Rev B 75:235330; Satpathi U, Singha Deo P (2012) IJMPB 26:1250028
2. Kalman O, Foldi P, Benedict MG, Peeters FM (2008) Physica E 40:567
3. Bagwell PF (1990) Phys Rev B 41:10354
4. Porod W, Shao Z, Lent CS (1993) Phys Rev B 48:8495 (R)
5. Singha Deo P et al (1998) Phys Rev B 58:10784
6. Bayman BF, Mehoke CJ (1983) Am J Phys 51(10):875

Chapter 5
Negative Partial Density of States

Abstract We first give an introduction to Argand diagram (AD) and Burgers circuit (BC). We then show that phase of the wave-function in different model systems can be understood in general using BC. This includes the new discontinuous phase change discussed in Chap. 2 that is understood as due to sub-loops in AD that is possible only in mesoscopic regime. The three prong potential is introduced and it is shown that the discontinuous phase change has continuous counterparts that lead to the dramatic effect that PDOS can become negative. Such continuous counterparts were not known in case of classical waves. FSR like rules can also be understood in terms of BC. General proofs are provided that make our results independent of the models used for demonstration. Finally, we re-interpret some existing experimental data with respect to Burgers circuit.

Fano resonance in quantum wires and associated phase drops as demonstrated in the previous Chap. 4 is of course well known in classical optics [1] but the mesoscopic phenomenon is different and novel. This difference is due to the distinct sub-band quantization (because of quantum confinement in y-direction as demonstrated in Fig. 2.8). Such a sub-band quantization has not yet been observed with optical wave guides essentially because of the much shorter wavelength of light. Besides phase changes are very important in mesoscopic physics as they are related to breakdown of parity effect, interpretation of Friedel sum rule (FSR), relation to partial density of states (PDOS), etc., that determine the thermodynamic properties of a mesoscopic system which is absent in case of optics. In this chapter we discuss the scattering phase shifts and their discontinuities from the point of view that is employed in optics and we will of course be motivated by parity effect and FSR like laws. This approach is related to the argument theorem in complex analysis and referred to as Burgers circuit. We will show that the Burgers circuit has a deep connection to density of states and partial density of states which was unknown before our works. This will help us prove the reality of negative PDOS and prove the general regimes when semi-classical FSR can be exact in a purely quantum regime. The advantage of using Burgers circuit is that these results can be shown to be very general [6] and thus goes beyond all earlier works.

P. S. Deo and K. Meena, *Time Travel: A Reality in Mesoscopic Physics*,
https://doi.org/10.1007/978-981-96-1136-2_5

We will show that the scattering phase shift behavior of different experimentally and theoretically studied systems can be understood from the Argand diagrams (AD) of scattering matrix elements. Different systems have their own peculiarities but it is possible to classify their behavior with respect to a mathematical principle like Burgers circuit. Argand diagram for the scattering matrix element of these systems can be classified as, (a) Argand diagram is closed, (b) Argand diagram is open, (c) Argand diagram encloses the phase singularity, (d) Argand diagram does not enclose the phase singularity, (e) Argand diagram is simply connected, and (f) Argand diagram is multiply connected due to the presence of sub-loops. Specific properties of the scatterer only matter to the extent that the Argand diagram changes from one of these to another. Among them some changes are topologically possible and others are not.

In Sect. 5.1 we will give an elementary introduction to Burgers circuit. In Sect. 5.2 we will analyze slips in the scattering phase shift for a few potentials. In Sect. 5.3 we will show that FSR too can be understood from Argand diagrams although its manifestation seems to be completely different for different potentials. We will also show why FSR can become exact whenever there is a phase lapse. This is a physically counter-intuitive result that has been proven for particular potentials so far. We will show that this result depends on the properties of Argand diagram and hence very general and independent of the scattering potential. In Sect. 5.3 we will also use Burgers circuit to prove the possibility of negative partial density of states in real mesoscopic systems. To show negative PDOS by explicit calculation for any particular realistic potential is virtually impossible and so never shown before. In Sect. 5.4 we present a brief summary of this chapter. In Sect. 5.5 we will present some experimental data and comment on them.

5.1 Burgers Circuit: An Introduction

Argand diagram is a plot of real versus imaginary parts of an analytic complex function and Burgers circuit is about phase changes and lapses being determined by phase singularities. Given a complex function t, AD is the plot of imaginary part of t or $Im(t)$ versus real part of t or $Re(t)$. The origin of this complex plane, that is the point $Re(t) = 0$, $Im(t) = 0$ is a phase singularity point. For a generalized "Burgers circuit" (BC) [M.V. Berry 1998],

$$\oint_C d\phi = 2\pi I \tag{5.1}$$

where $\phi = Arctan\frac{Im(t)}{Re(t)}$, which makes it clear that the origin in the AD plot is a singularity point for ϕ. I is a topological quantum number which is always conserved in an interaction. I is a sign and so can be $+1$, -1, and 0. If the contour C does not enclose the phase singularity then I is 0. When the contour C enclosing a phase singularity is clockwise then I is -1 and when the contour C enclosing a phase singularity is counter-clockwise then I is +1. A scattering matrix element is a complex function for which Argand diagram can be drawn and concept of BC can be applied. For the rest of this chapter where we refer to scattering matrix element, we will mean the scattering matrix element of a quantum mechanical particle say an electron. It is generally seen that AD for a scattering matrix element for any given potential is counter-clockwise, as incident energy of the scattering particle increases. Figure 5.1a and b shows imitation Argand diagrams for a complex function t that will help understand real ones. Figure 5.1a shows a typical counter-clockwise contour (ABDEA) of an Argand diagram enclosing the phase singularity at the origin. Hence for this case, $I = +1$. The contour trajectory (ABDEA) is concave throughout with respect to the singular point at the origin. Following Eq. (5.1), the net change in phase in tracing ABDEA, in Fig. 5.1a, is 2π. Thus the phase monotonously increases for a concave trajectory. One can easily understand this by taking a point on the AD and calculate $\phi = Arctan\frac{Im(t)}{Re(t)}$ at that point, where obviously $\left(Re(t), Im(t)\right)$ is the coordinates of that point. The figures show some arrows schematically representing the ϕ at those points. Figure 5.1b shows a typical contour (FGHJF) of an Argand diagram not enclosing the phase singularity at the origin. Hence for this case, $I = 0$. The contour has both concave (FGH) and convex (HJF) trajectories with respect to the singular point at the origin. Following Eq. (5.1), the total change in phase in Fig. 5.1b is zero. It is possible if the phase increases for the concave trajectory and decreases for the convex trajectory, the net phase change being zero. This again can be easily verified by calculating ϕ at any point $\left(Re(t), Im(t)\right)$ of the trajectory and will be further demonstrated below. Thus for both Fig. 5.1a and b, the phase change is determined by the singular point and the two follow the same principle. As a special case of the two, one can have a situation where the contour touches the singular point. In which case too the phase change can be understood from the same principle and will be explained later. Also if it is a closed contour as in Fig. 5.1a and b, then Eq. (5.1) is exact. In the real systems we will discuss in this work, the contours may not be closed. However, we will extend Eq. (5.1) to understand such cases too.

Whatever be the shape of a closed contour C, the phase change is given by Eq. (5.1). As a consequence, the real and imaginary parts of t are not independent of each other but are related. Let us say the complex transmission amplitude is

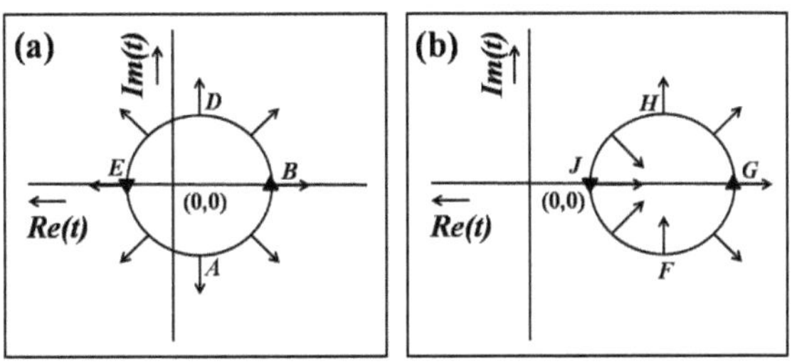

Fig. 5.1 Schematic Argand diagrams that exemplify Eq. 5.1. **a** When the Argand diagram contour encloses the singular point at (0, 0) and **b** when the contour does not enclose the singular point

$t(U) = |t(U)|e^{i\theta_t(U)}$ where U can be any parameter like incident energy or gate voltage. Then

$$Re\,(t(U)) = \frac{1}{\pi}P\int_{-\infty}^{\infty}\frac{Im(t(U'))}{U'-U}dU' \tag{5.2}$$

$$Im\,(t(U)) = -\frac{1}{\pi}P\int_{-\infty}^{\infty}\frac{Re(t(U'))}{U'-U}dU' \tag{5.3}$$

These are the well known Kramers-Kronig relations. Another way in which the relation can be stated is in terms of Hilbert Transform,

$$ln|t(U)| = \frac{1}{\pi}P\int_{-\infty}^{\infty}\frac{\phi(U')}{U'-U}dU' \tag{5.4}$$

$$\phi(U) = -\frac{1}{\pi}P\int_{-\infty}^{\infty}\frac{ln|t(U')|}{U'-U}dU' \tag{5.5}$$

Since I is a conserved quantity, by adding terms to a Hamiltonian (or details to the states in the scatterer) we cannot remove the phase singularities of the wavefunction in the complex plane. Phase changes are determined by the phase singularities. Depending on the interaction the shape of the Argand diagram can however change and so a theoretical understanding of the experimental data may not crucially depend on the details of the sample or model. Sample details can change the shape of the contour C, but as these theorems state, to understand the phase changes, we do not need all these details. Englman and Yahalom [2] had shown that the experimental data for scattering phase shift and scattering cross section of a quantum dot are consistent with Hilbert transforms. We will show that the principles of analyticity and Eq. (5.1) can be used to establish our results rigorously [6].

5.2 Model Potentials

In this section we analyze scattering phase shifts for the different potentials with respect to Eq. (5.1). As will be clear, we do not need very complicated realistic potentials to understand the phase shifts but we need representative potentials that can be exactly solved and help us understand different aspects of Eq. (5.1).

5.2.1 Double Delta Function Potential in One Dimension

Let us first consider scattering by a double delta function potential in one dimension (1D) schematically shown in Fig. 5.2a. Although a simple potential, it exhibits pronounced Breit-Wigner (BW) resonances. We will use Eq. (5.1) to understand the scattering phase shift for this system and hence for BW resonances. The scattering potential for this system can be written as

$$U_1(x) = \gamma_1 \delta(x)$$

$$U_2(x) = \gamma_2 \delta(x - a)$$

where γ_1 and γ_2 are the strengths of potential U_1 and U_2, respectively. The wave-functions in different regions marked I, II, and III are (see Fig. 5.2a)

$$\psi(x) = \begin{cases} e^{ikx} + re^{-ikx}, \; for\; x < 0, \\ Ae^{ikx} + Be^{-ikx}, \; for\; 0 \leq x \leq a, \\ te^{ik(x-a)}, \; for\; x > a \end{cases}$$

Here r and t are the reflection and transmission amplitudes, $k = \sqrt{\frac{2m_0}{\hbar^2} E}$ is the wave vector, E is the incident Fermi energy, $t =\mid t \mid e^{i\theta_t}$, $\theta_t = Arctan\frac{Im(t)}{Re(t)}$ is transmission phase shift and $|t| = \sqrt{Im(t)^2 + Re(t)^2}$ is transmission modulus. The Argand diagram for t is shown in Fig. 5.2b, where energy is varied to remain within the first Riemann surface. There is a phase singularity at the origin where $t = 0$. The Argand diagram encloses the singularity but is not closed in the first Riemann surface. Figure 5.2c and d shows the transmission phase shift θ_t and the transmission coefficient $|t|^2$, respectively, as a function of ka, using the same parameters as in Fig. 5.2b.

In Fig. 5.2b, the contour starts from the origin where $E = 0$, goes first through point P, and then through Q, R, and S. The trajectory facing the singular point at the origin is concave throughout, and thus as discussed with Fig. 5.1, the phase increases continuously. This is evident in Fig. 5.2c where the points P, Q, R, and S are also shown at their respective values of ka. As the trajectory comes closer to the point of phase singularity, phase changes are very small and energy cost is very high. In Fig. 5.2b, the energy at the points marked P, Q, R, and S are $8.614a$, $9.12a$, $23.61a$, and $37.58a$. Thus the energy change in going from P to Q (a large arc in the trajectory in Fig. 5.2b) is very small, whereas the energy change in going from Q to R (a small arc in the trajectory in Fig. 5.2b) is very high. The point R is very close to the singular point. Thus it costs a lot of energy as the Argand diagram trajectory tries to approach the point of phase singularity.

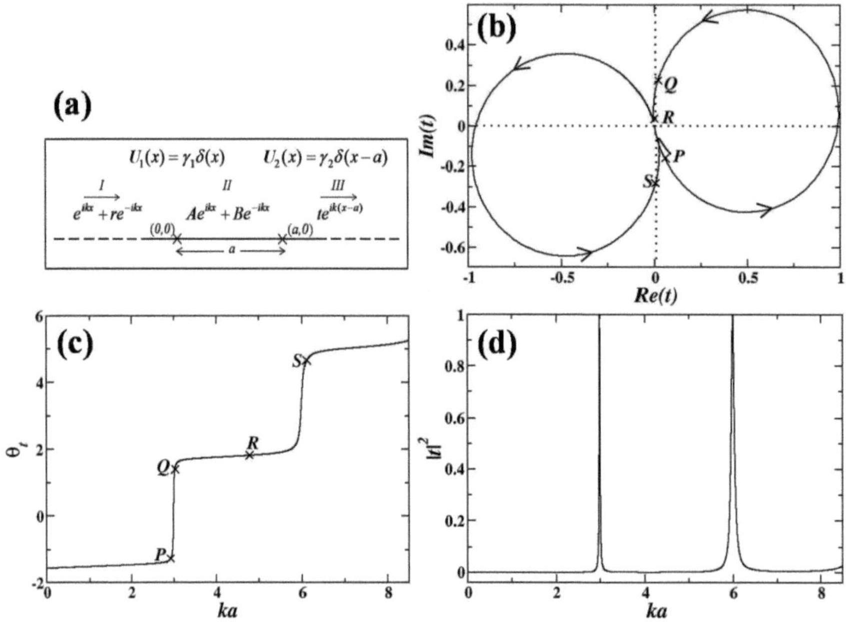

Fig. 5.2 **a** Schematic representation for scattering of electrons by a double delta function potential in one dimension. The direction of incident and transmitted electrons are represented by arrows. The solid line represents a quantum wire with double delta function potentials at positions $x = 0$ and $x = a$, respectively, shown by cross (X) marks. γ_1 and γ_2 are the strengths of the potentials. The dashed lines represent the fact that the quantum wire is connected to electron reservoirs via leads. **b** Argand diagram for transmission amplitude for the double delta function potential. **c** Plot of transmission phase shift θ_t versus ka and **d** plot of transmission coefficient $|t|^2$ versus ka, for the double delta function potential using parameters $e_0\gamma_1 a = e_0\gamma_2 a = 40$, $a = 1$, $e_0 = 1$, $2m_0 = 1$, and $\hbar = 1$

5.2.2 Stub Potential

A slightly modified form of the stub is schematically shown in Fig. 5.3a. We will use Eq. (5.1) to understand why scattering phase shift for this system can change discontinuously as some parameter is varied. Electrons are incident from left (see Fig. 5.3a) with energy E. The thin lines represent one-dimensional quantum wires with zero potential, while the bold lines represent quantum wires with a finite potential $U(y)$ given by

$$U(y) = \begin{cases} 0, \textit{for}\ 0 < y \leq l_1, \\ iU_0, \textit{for}\ l_1 < y < l, \\ \infty, \textit{for}\ y \geq l \end{cases}$$

The potential $U(y)$ is taken to be imaginary, as it allows us to make the Argand diagram trajectory approach and cross the point of phase singularity. This cannot

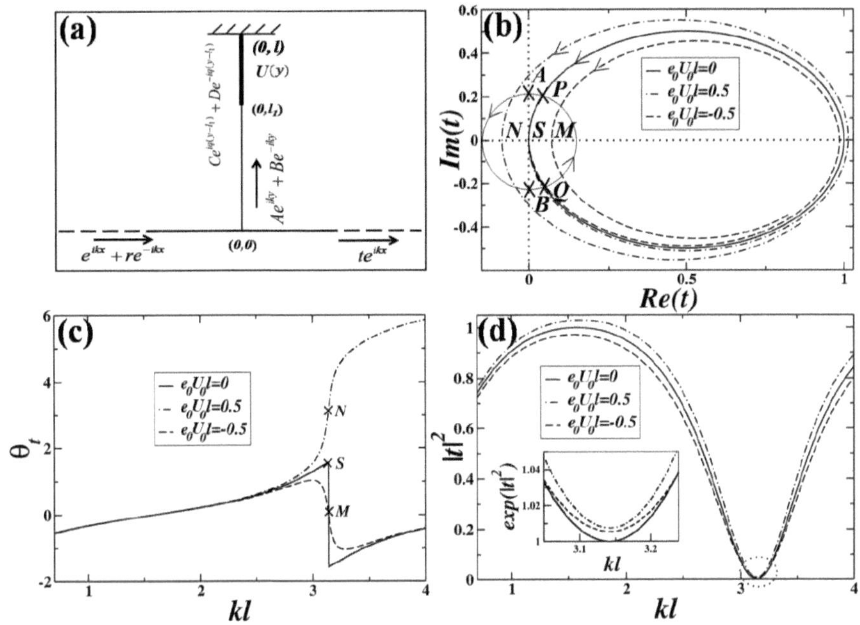

Fig. 5.3 **a** Schematic representation for scattering of electrons by a stub potential. The direction of incident and transmitted electrons are shown by arrows. The potential, represented by the bold line along y-axis, is $U(y) = iU_0$. **b** Argand diagram for transmission amplitude for different values of e_0U_0l, where e_0 is electronic charge and l is the length shown in **a**. The thick solid line is Argand diagram for the case when $e_0U_0l = 0$, the dot-dashed line is that for $e_0U_0l = 0.5$, and the dashed line is for $e_0U_0l = -0.5$. Here $l_1 = .5l$, $l = 1$, $e_0 = 1$, $2m_0 = 1$, and $\hbar = 1$. **c** Plot of transmission phase shift θ_t and **d** plot of transmission coefficient $|t|^2$, as a function of dimensionless wave vector kl, taking the same notations and parameters as in Fig. 5.3b

be done with real potentials as I in Eq. (5.1) is a conserved quantity. Imaginary potentials are known as optical potentials [3]. They are often used to simulate the effect of decoherence and open systems. If the system changes from a closed one to an open one then I may be different for the two cases. When U_0 is 0 then $U(y)$ is 0 and the system is a closed conserved system. And when U_0 is non-zero then $U(y) = iU_0$ and it is an open system. The wave-functions in the different regions are given by

$$\psi(x, y) = \begin{cases} e^{ikx} + re^{-ikx}, for\ x < 0, \\ te^{ikx}, for\ x > 0, \\ Ae^{iky} + Be^{-iky}, for\ 0 < y \leq l_1, \\ Ce^{iq(y-l_1)} + De^{-iq(y-l_1)}, for\ l_1 < y < l, \\ 0, for\ y = l \end{cases}$$

Here r and t are the reflection and transmission amplitudes, $k = \sqrt{\frac{2m_0}{\hbar^2}E}$ is the wave vector along thin lines, $q = \sqrt{\frac{2m_0}{\hbar^2}(E - iU_0)}$ is the wave vector along the bold line, and E is the Fermi energy. Solving the scattering problem using boundary conditions elaborated in Chap. 1, for Fig. 5.3a we get r and t as a function of energy E.

$$r = \frac{1 - \frac{ik-q_1}{ik+q_1}e^{-2ikl_1}}{1 + 3\frac{ik-q_1}{ik+q_1}e^{-2ikl_1}} \tag{5.6}$$

$$t = \frac{2(1 + \frac{ik-q_1}{ik+q_1}e^{-2ikl_1})}{1 + 3\frac{ik-q_1}{ik+q_1}e^{-2ikl_1}} \tag{5.7}$$

where $q_1 = q\,[cot\,(q(l - l_1))]$. Transmission phase shift is given by $\theta_t = Arctan\frac{Im(t)}{Re(t)}$. Figure 5.3b shows the Argand diagrams for the transmission amplitudes, given by Eq. (5.7), for different values of e_0U_0l, where energy is varied to remain within the first Riemann surface. The thick solid line is for $e_0U_0l = 0$, the dot-dashed line is for $e_0U_0l = 0.5$, and the dashed line is for $e_0U_0l = -0.5$. There is a phase singularity at $t = 0$ marked as S in Fig. 5.3b. Figure 5.3c and d shows the transmission phase shift θ_t and transmission coefficient $|t|^2$, respectively, as a function of kl for different values of e_0U_0l, using the same parameters and same notations as in Fig. 5.3b.

In Fig. 5.3b, if we draw a circle (thin solid line through $PABQP$) around the phase singularity, then Eq. (5.1) implies the total phase change along this contour is 2π. If the shape and size of the contour are altered, the net phase change along the contour remains the same provided the contour encircles the phase singularity. Now follow the thick solid curve in Fig. 5.2b. The phase change while going from P to Q along the arrow in the contour is less than π as the phase change between the points A and B is π. In the limit when the radius of the circle $PABQ$ is tending to zero, P approaches A, Q approaches B, and P, A, B, Q all coincide with the singular point S. In this limit, going from P to Q along the thick solid line would imply a discontinuous phase change of π. Therefore a trajectory that tangentially touches the singular point S like the thick solid line for $e_0U_0l = 0$ in Fig. 5.3b will exhibit a discontinuous phase change of π at S. This phase change can be seen in the solid line for $e_0U_0l = 0$ in Fig. 5.3c (at point S marked at the same value of kl as in Fig. 5.3b), where we have plotted the transmission phase shift θ_t versus kl. We have already discussed in the previous section that as the Argand diagram trajectory approaches the point of phase singularity, the energy cost is very high. So, it is surprising that in a mesoscopic system, one can have such a high energy scale and one can realize discontinuous phase drops that are generally not observed for classical wave transport. In fact, it is not a real energy scale but an effective energy scale. Scattering by a stub of length l can be mapped into a problem of a delta function potential in one dimension where the strength of the delta function potential is $kcot(kl)$. That is, $U^{eff}(x) = kCot(kl)\delta(x)$ has same reflection amplitude r and transmission amplitude t as a stub of length l. This is an effective potential that the electrons encounter while the real potential is $U_0 = 0$. At $kl = \pi$, a small change in k or l means a very large change in the effective potential $U^{eff}(x)$. Such large effective energy scales are known in other areas of condensed matter physics, for example, the effective electron mass becomes ∞ at the band edge.

For $e_0 U_0 l = 0.5$, the Argand diagram trajectory shown by the dot-dashed line in Fig. 5.3b intersects the $Re(t)$ axis at point N at $kl = \pi$. The Argand diagram trajectory facing the phase singularity at S is concave throughout. So, Eq. (5.1) implies a monotonously increasing phase. This phase behavior can be seen in the dot-dashed line in Fig. 5.3c where we have plotted the transmission phase shift θ_t versus kl and marked N at $kl = \pi$. For $e_0 U_0 l = -0.5$, the Argand diagram trajectory shown by the dashed line in Fig. 5.3b intersects the $Re(t)$ axis at point M at $kl = \pi$. The Argand diagram trajectory facing the phase singularity is partially convex and partially concave. As discussed with respect to Eq. (5.1), the phase increases for the concave part and decreases for the convex part of the trajectory. This phase behavior can be seen in the dashed curve in Fig. 5.3c where we have plotted the transmission phase shift θ_t versus kl and marked the point M similarly. Therefore, if $e_0 U_0 l$ is continuously changed from 0.5 to -0.5, then the point N moves to M crossing the point of phase singularity. When $e_0 U_0 l$ becomes 0 we get the solid curves in Fig. 5.3b, c, and d. This implies that with an imaginary potential, just by changing sign of the potential it is possible to cross the singular point. I for dot-dashed line is 1 and that for the dashed line is 0. In other words by changing an imaginary potential we can make I change from 1 to 0. This is difficult with a real potential. If we tried to cross the singular point with real potentials it would cost an infinite amount of energy. We have thus discussed an effective real potential that can make an Argand diagram trajectory tangentially touch the point of phase singularity. Any real effective potential that can make Argand diagram trajectory cross the point of phase singularity is not known.

In Fig. 5.3c, we observe two different types of phase drops. For $e_0 U_0 l = 0$, we get a discontinuous phase drop (shown by the solid curve in Fig. 5.3c) and for $e_0 U_0 l = -0.5$, we get a gradual phase drop (shown by the dashed curve in Fig. 5.3c). If we make $e_0 U_0 l > -0.5$, the point M in Fig. 5.3b will shift more to the right and the phase drop will be less in magnitude and also less sharp. In Fig. 5.3d, using the same parameters and same notations as in Fig. 5.3b and c, transmission coefficient $|t|^2$ is plotted as a function of kl for different values of $e_0 U_0 l$. The inset in Fig. 5.3d shows the transmission coefficient $|t|^2$ in exponential scale in the region around $kl = \pi$ (shown by dotted circle). At $kl = \pi$, the thick solid curve for $e_0 U_0 l = 0$ goes to zero, while the dot-dashed and dashed curves for $e_0 U_0 l = 0.5$ and $e_0 U_0 l = -0.5$, respectively, go through a non-zero minima and are very close. The phase behaviors for dot-dashed and dashed curves are completely different as can be seen from Fig. 5.3c.

Thus in this section, we have explained using Argand diagram and Burgers circuit, why phase (scattering phase shift) drops can occur discontinuously? Why such a phase drop can disappear or change from discontinuous to gradual? The gradual phase drop in the dashed curve of Fig. 5.3c at M is related to the analytic property of complex transmission amplitude and how the trajectory encloses the singularity.

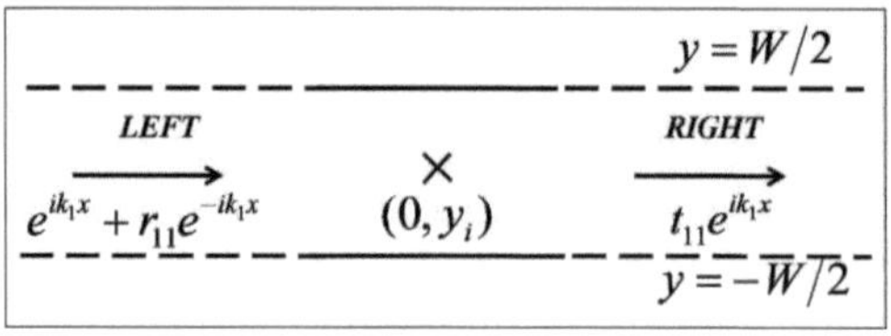

Fig. 5.4 Schematic representation for scattering of electrons by a delta function potential in quasi-one dimension. The position of a delta function potential is shown by cross (X) mark

Whether the gradual drop is sharp or slow depends on the distance SM (Fig. 5.3b) at which the convex trajectory intercepts the $Re(t)$ axis.

5.2.3 Delta Function Potential in Q1D

The next scattering potential we consider is a delta function potential in a single channel quantum wire. This essentially means there is a single propagating channel while all other channels are evanescent as elaborated in Chap. 4. These evanescent channels are characteristic of quasi-one dimension (Q1D) and make this scattering potential completely different from that of a delta function potential in one dimension. This system gained relevance with respect to some recent experiments [7, 8]. The scattering solution for this potential has been discussed in detail in Sect. 4.2 and so we directly proceed to analyze the scattering phase shift in terms of Argand diagrams.

The Argand diagram for transmission amplitude t_{11} is shown in Fig. 5.5a. There is a phase singularity at the origin where $t_{11} = 0$ (shown by the point S). Figure 5.5b shows transmission phase shift $\theta_{t_{11}}$ as a function of energy, for the same parameters as in Fig. 5.5a. In Fig. 5.5a, the trajectory (shown by thick solid line) starts from the origin (point marked S), goes up to point Q, and traces back the same path to pass

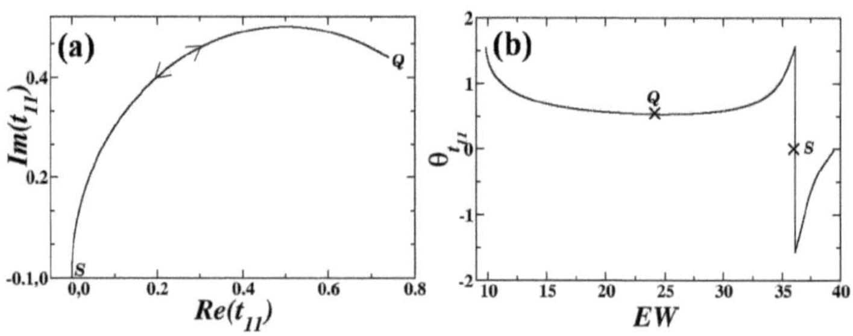

Fig. 5.5 **a** Argand diagram for transmission amplitude and **b** plot of transmission phase shift $\theta_{t_{11}}$ versus EW, for scattering by a delta function potential in a single channel quantum wire of width W shown in Fig. 5.4. Here $e_0\gamma W = -1.5$, $y_i = 0.21W$, $e_0 = 1$, $W = 1$ and we have considered 500 evanescent modes

the origin making SQS a closed contour. The direction of the trajectory is therefore shown by a double headed arrow. At the energy, where the trajectory goes from Q to S and touches the point of phase singularity at origin, i.e., point S, we expect a discontinuous phase drop of π, following the same argument as in the case of the stub. This phase drop can be seen in Fig. 5.5b, where the points Q and S are also marked at their respective energies. Thus the discontinuous phase drop is a natural consequence of Eq. (5.1).

5.2.4 Three Prong Potential

We now consider another potential called the three prong potential. This potential will help us to demonstrate other non-trivial aspects that follow from Eq. (5.1). A schematic representation of the three prong potential is shown in Fig. 5.6. The thin lines represent one-dimensional quantum wires with potential $U = 0$, and the bold lines represent quantum wires with non-zero potential, i.e., $U \neq 0$. The arms of the prong are labeled as 1, 2, and 3 as shown in Fig. 5.6. The electrons are considered to be incident from left, the direction of incidence being shown by arrows. The wave-functions in the different regions are given by

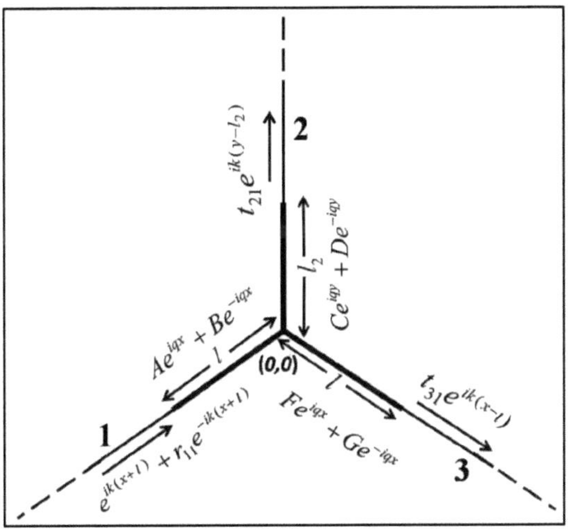

Fig. 5.6 Schematic representation of scattering of electrons by a three prong potential. The directions of incident and transmitted electrons are shown by arrow heads. The potential is non-zero along the bold lines of lengths l, l_2, and l along $-x$, $+y$, and $+x$ axes, respectively

$$\psi(x, y, z) = \begin{cases} e^{ik(x+l)} + r_{11}e^{-ik(x+l)}, for\ x < -l, \\ Ae^{iqx} + Be^{-iqx}, for\ -l < x < 0, \\ Ce^{iqy} + De^{-iqy}, for\ 0 < y < l_2, \\ Fe^{iqx} + Ge^{-iqx}, for\ 0 < x < l, \\ t_{21}e^{ik(y-l_2)}, for\ y > l_2, \\ t_{31}e^{ik(x-l)}, for\ x > l. \end{cases}$$

where $k = \sqrt{\frac{2m_0}{\hbar^2}E}$ is the wave vector along the thin lines, $q = \sqrt{\frac{2m_0}{\hbar^2}(E-U)}$ is the wave vector along the bold lines, and E is the Fermi energy. Here r_{11} is the reflection amplitude for electrons incident from 1 and reflected back to 1, t_{21} is the transmission amplitude for electrons incident from 1 and transmitted to 2, and t_{31} is the transmission amplitude for electrons incident from 1 and transmitted to 3. These scattering matrix elements can be solved using boundary conditions elaborated in Chap. 1. The respective transmission phase shifts are given by $\theta_{r_{11}} = Arctan\frac{Im(r_{11})}{Re(r_{11})}$, $\theta_{t_{21}} = Arctan\frac{Im(t_{21})}{Re(t_{21})}$, and $\theta_{t_{31}} = Arctan\frac{Im(t_{31})}{Re(t_{31})}$.

Figure 5.7a shows the Argand diagram for transmission amplitude t_{21} . There is a phase singularity at the origin, where $t_{21} = 0$. In Fig. 5.7a the trajectory of Argand diagram for t_{21} starts from the origin, goes through P, and then through Q, R, and S following counter-clockwise direction shown by arrows. The trajectory is concave throughout and Eq. (5.1) implies monotonously increasing phase. This monotonously increasing phase can be seen in Fig. 5.7b, where scattering phase shift $\theta_{t_{21}}$ is plotted as a function of kl and here also the points P, Q, R, and S are marked at the corresponding values of kl.

The Argand diagram for t_{31} shows something interesting. This Argand diagram is shown in Fig. 5.7c. There is a phase singularity at the origin, where $t_{31} = 0$. In Fig. 5.7c, the trajectory of Argand diagram for t_{31} starts from origin, goes through P, and then through D, A, B, Q, F, A, and S, following counter-clockwise direction shown by arrows. Here interestingly, the trajectory develops a sub-loop $ABQFA$. This sub-loop results in a convex arc BQF in the trajectory that does not go through the origin. As explained earlier, there will be a gradual phase drop whenever such a convex arc is observed, following Eq. (5.1). This can be seen in Fig. 5.7d, where scattering phase shift $\theta_{t_{31}}$ is plotted as a function of kl and here also the points P, A, Q, R, and S are marked. Presence or absence of such a sub-loop has no consequence on the line integral of phase along $PDABQFARS$. This is because the contribution to the line integral coming from the sub-loop $ABQFA$ is 0 and its presence or absence has no bearing on the value of I. So, such a sub-loop as $ABQFA$ in Fig. 5.7c can appear or disappear (in the sense that it can reduce to a cusp) as some parameter is varied as will be demonstrated in the next section.

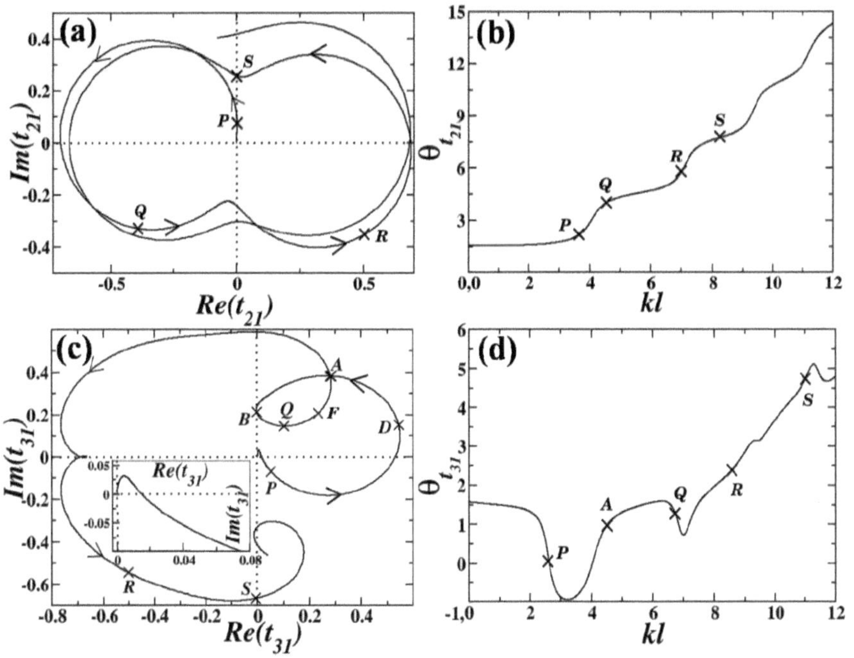

Fig. 5.7 The system in consideration is shown in Fig. 5.6. **a** Argand diagram for transmission amplitude t_{21} and **b** plot of scattering phase shift $\theta_{t_{21}}$ as a function of kl. **c** Argand diagram for transmission amplitude t_{31} and **d** plot of scattering phase shift $\theta_{t_{31}}$ as a function of kl. For all the figures, $l = 1$, $l_2 = 5l$, $e_0 = 1$, and $e_0 U l = -1000$

5.3 Injectance and Friedel Sum Rule

Note substituting $s_{\alpha\beta} =| s_{\alpha\beta} | e^{i\theta_{s_{\alpha\beta}}}$, PDOS in Eq. 3.46 can be written as

$$\rho_{pd}(E, \alpha, \beta) = -\frac{1}{2\pi} \int_{sample} d^3\mathbf{r} \left(|s_{\alpha\beta}|^2 \frac{\delta \theta_{s_{\alpha\beta}}}{\delta U(\mathbf{r})} \right) \tag{5.8}$$

Here, the region of integration is the "sample" which was earlier denoted by Ω, $s_{\alpha\beta}$ is the scattering matrix element for electrons incident from channel β and transmitted to channel α, and $\frac{\delta}{\delta U(\mathbf{r})}$ stands for a functional derivative with respect to the local potential $U(\mathbf{r})$. Electronic charge e_0 is set to 1.

It is to be noted that the LHS of Eq. 5.8 gives the object of study and the RHS of Eq. 5.8 is an exact mathematical expression for the object. Although there is an equality between the two, a mathematician will say that an object and its mathematical expression are different entities. This of course is a philosophical point of view that is often framed as the grand fathers axe paradox, which shows it is very difficult to define the object grandfather's axe. Note that is different from the grandfather's paradox. The grandfather and his axe pose two different paradoxes. A mathematical expression for an object is similar or dissimilar with the measured value of the object in the sense that an experimentalist can completely deny the expression but obviously accept the object because that is what is measured with laboratory equipments. Likewise, a mathematician is often stubborn enough to ignore measurements. Mathematicians often take pride in defining abstract objects that have nothing to do with nature and measurements. One can always ignore this philosophical tug-of-war, except for the fact that the axiomatic framework of quantum mechanics denies the existence of the object itself. When we use terms like LPDOS or ρ_{pd} or LPT, then we are referring to the object. In this section we will essentially justify the expression for the object. While in Chap. 7, we will justify the measurability of the object (see Eq. 7.13 for example where the LHS is what an experimentalist will measure independent of any theory). Both can be justified from AD and BC as a winding number, which should convince both parties, that is the mathematician as well as the experimentalist, that the object in terms of an expression or in terms of measurability is just a consequence of the law of numbers, which both parties accept. That in turn should convince both, the reality of time travel, which will be discussed in Chap. 6. The essential argument is that in this chapter we will identify ρ_{pd} as a winding number in the complex plane of the relevant scattering matrix element and a winding number can be an integer, or a fraction represented by a real number, positive as well as negative. Quantitative justification will be provided in Chap. 6.

The integration over $\mathbf{r}$ in Eq. 5.8 can easily be done for a global change (for all $\mathbf{r}$ in the sample as well as in the leads) in $U(\mathbf{r})$ by a constant amount ϵ, i.e., $\delta U(\mathbf{r}) = \epsilon$ for all $\mathbf{r}$. Such a constant global increase in potential is equivalent to decrease in incident energy E, i.e.,

$$\int_{global} d^3\mathbf{r} \frac{\delta}{\delta U(\mathbf{r})} \equiv -\frac{d}{dE} \tag{5.9}$$

and, therefore,

$$\int_{sample} d^3\mathbf{r} \frac{\delta}{\delta U(\mathbf{r})} \cong -\frac{d}{dE} \tag{5.10}$$

is expected to work in the semi-classical limit. So, from Eq. 5.8 or 3.10

$$\begin{aligned}
\rho_{pd}(E,\alpha,\beta) &= -\frac{1}{4\pi i}\int_{sample} d^3\mathbf{r}\left(s^*_{\alpha\beta}\frac{\delta s_{\alpha\beta}}{\delta U(\mathbf{r})} - s_{\alpha\beta}\frac{\delta s^*_{\alpha\beta}}{\delta U(\mathbf{r})}\right)\\
&\approx -\frac{1}{4\pi i}\int_{global} d^3\mathbf{r}\left(s^*_{\alpha\beta}\frac{\delta s_{\alpha\beta}}{\delta U(\mathbf{r})} - s_{\alpha\beta}\frac{\delta s^*_{\alpha\beta}}{\delta U(\mathbf{r})}\right)\\
\text{or,}\quad \rho_{pd}(E,\alpha,\beta) &\approx \frac{1}{4\pi i}\left(s^*_{\alpha\beta}\frac{ds_{\alpha\beta}}{dE} - s_{\alpha\beta}\frac{ds^*_{\alpha\beta}}{dE}\right) \qquad \text{(Using Eq. (5.9))}
\end{aligned}$$

On substituting $s_{\alpha\beta} = \mid s_{\alpha\beta} \mid e^{i\theta_{s_{\alpha\beta}}}$ we get

$$\rho_{pd}(E, \alpha, \beta) \approx \frac{1}{2\pi}\left(|s_{\alpha\beta}|^2 \frac{d\theta_{s_{\alpha\beta}}}{dE}\right) \tag{5.11}$$

We have already shown that in particular cases $\frac{d\theta_{s_{\alpha\beta}}}{dE}$ can be negative. However, concluding $\rho_{pd}(E, \alpha, \beta)$ as given by the RHS of Eq. 5.8 to be negative when R.H.S of Eq. (5.11) is negative is completely wrong. That is because Eq. (5.11) is expected to be an approximate equality, and for mesoscopic systems it is not even correct to say a semi-classical approximation as we will show. For Eq. 5.11, if R.H.S is negative, the L.H.S, as given by Eq. 5.8, is not necessarily negative and may well be strongly positive. It is also generally not possible to numerically verify the negativity of the expression in RHS of Eq. 5.8 because in 1D as we will show it does not become negative (see Fig. 6.3 and the associated discussions) and in Q1D or more it is technically and conceptually [5] very difficult. We do not refer to cases like the three prong potential as 1D but 1D refers to the typical square well potential and analogues of it. The conceptual difficulties are associated with the fact that quantum mechanics do not allow us to address a local partial state of a particle unless some ensemble averaging is allowed as explained in Sect. 3.3. However, RHS of Eq. 5.8 is well defined as we have justified in Chap. 3, and also corresponds to observables as we will show in the last chapter. We will show that some general conclusions can be drawn about the R.H.S of Eq. 5.8 using the properties of Argand diagram and Burgers circuit which is true for any potential in 1D or in Q1D. For any potential that can support Fano resonances the RHS of Eq. 5.8 can become negative at the Fano resonance. The three prong potential and attractive potentials in Q1D are examples of it. Note that although we have claimed that Eq. 5.10 does not give the semi-classical limit, we will refer to it as a semi-classical approximation as in standard literature and the reader has to understand from the graphs and discussions, when it breaks down and when not.

Even though PDOS as expressed in Eq. 5.8 can be negative as we will show, they add up to give the correct DOS which is positive. We can sum PDOS to get injectance and DOS as we have shown in Chap. 3. Using Eq. (5.10) we get the semi-classical limit of Eq. 3.49, which gives semi-classical injectance as

$$\rho(E, \beta) \approx \frac{1}{4\pi i}\sum_{\alpha}\left(s^*_{\alpha\beta}\frac{ds_{\alpha\beta}}{dE} - s_{\alpha\beta}\frac{ds^*_{\alpha\beta}}{dE}\right) \tag{5.12}$$

Summing $\rho(E, \beta)$ over the β independent channels, we can obtain density of states (DOS) $\rho_d(E)$, i.e., from Eq. 3.51,

$$\rho_d(E) = -\frac{1}{4\pi i}\sum_{\alpha\beta}\int_{sample} d\mathbf{r}^3\left(s^*_{\alpha\beta}\frac{\delta s_{\alpha\beta}}{\delta U(\mathbf{r})} - s_{\alpha\beta}\frac{\delta s^*_{\alpha\beta}}{\delta U(\mathbf{r})}\right) \tag{5.13}$$

and in the semi-classical limit given by Eq. (5.10), we get

$$\rho_d(E) \approx \frac{1}{4\pi i} \sum_{\alpha\beta} \left(s^*_{\alpha\beta} \frac{ds_{\alpha\beta}}{dE} - s_{\alpha\beta} \frac{ds^*_{\alpha\beta}}{dE} \right) \tag{5.14}$$

Further simplification of R.H.S in Eq. (5.14) gives

$$\pi \rho_d(E) \approx \frac{d}{dE} \theta_f(E) \tag{5.15}$$

This is Friedel sum rule (FSR), where $\theta_f(E) = \frac{1}{2i} log(det[S])$ is the Friedel phase, S is the scattering matrix, and $\rho_d(E) = \frac{dn(E)}{dE}$ is density of states. Since injectance of all leads are independent (as shown in Sect. 1.1) while they add up to give DOS, it is important to understand injectance in order to understand FSR. So we will restrict our study to injectance.

As we have shown in Chap. 4 semi-classical regime being expressed by Eq. (5.10) does not seem to be sufficient. Sometimes, FSR (or injectance) is exact at all energies (for example the stub) and sometimes it is exact at an energy where quantum fluctuations dominate. There is a huge amount of system to system variation. We have also proved very generally in Chap. 4 that when the phase drops of π are discontinuous like that in the solid curve in Fig. 5.3c or Fig. 5.5b, then at the energy corresponding to this drop, semi-classical injectance will become exact. But when the phase drops are gradual like the dashed curve in Fig. 5.3c, this has not been proved in general but shown for the particular case of a delta function potential in Q1D. Burgers circuit will help us to derive a general proof that gradual phase drops imply exactness of semi-classical injectance for any arbitrary potential that can exhibit such a gradual phase drop. We will first prove that in realistic mesoscopic systems one can definitely observe negative PDOS. We will show that the R.H.S of Eq. (5.8) can become negative for a realistic mesoscopic system. We will also show that such a conclusion cannot be drawn in 1D that has been extensively studied before [5]. Also we will show that when there are such negative slopes in scattering phase shift of mesoscopic systems then semi-classical FSR can become exact in a quantum regime. Once again our proofs will depend on Argand diagram and Eq. (5.1) and so our proof will be general and not depend on the specific properties of the scattering potential.

For this we first make the connection between Eq. (5.1) and injectance. When S matrix is 2×2 (for example, the cases of double delta function potential of Fig. 5.2a, stub of Fig. 5.3a, single channel quantum wire of Fig. 5.4), Eq. (5.15) simplifies to

$$\frac{d}{dE} \theta_t(E) \approx \pi \rho_d(E) \tag{5.16}$$

with $\rho_d(E) = \frac{dn}{dE}$. Here t is the transmission amplitude and $\theta_t = Arctan \frac{Im(t)}{Re(t)}$. It has been proved in Chap. 4 that this equation is valid even if θ_t is discontinuous with

E because θ_t is analytic wherein the right derivative of θ_t is the same as the left derivative at the discontinuity.

Suppose when the energy is varied from 0 to E_1, then the Argand diagram for a typical scattering matrix element t traces a closed contour C. Below we will show that for such a case one can state that Eq. (5.1) takes the form

$$\oint_C d\theta_t = \pi N(E_1) \tag{5.17}$$

where $N(E_1)$ is number of states (obtained by integrating DOS $\rho_d(E)$ from 0 to E_1) below energy E_1 and is to be identified with the conserved quantity I (see Fig. 5.2b for the factor 2). We start with intuitive arguments before proving this in general. Comparing with Eq. (5.1) we see $\phi \equiv \theta_t$ and $2I \equiv N(E_1)$. If C happens to be a completely closed contour, Eq. (5.17) turns out to be exact and equivalent to Eq. (5.1). Any complex function or its phase has to satisfy Eq. (5.1) and a scattering matrix element is no exception provided it is analytic. This analyticity is the basic requirement for Eq. (5.8) (PDOS) or Eq. 3.49 (injectance). However, in most cases of scattering problems, C is not completely closed and when C is not completely closed one cannot expect any conserved quantity. In Fig. 5.2b, there is a phase singularity at the origin and the contour enclosing the singularity is not closed in the first Riemann surface. One can restrict the discussion to the first Riemann surface to understand the injectance. When the contour continues to the second Riemann surface, then the contour integral starts including contribution from the second phase singularity in the second Riemann surface. And then one has to extend the discussions here to include the effect of the second singularity. This does not change the arguments given here except that sometimes the error from the first Riemann surface can cancel the error from the second Riemann surface which need not be a systematic behavior. The contour C in Fig. 5.2b starts from origin with zero energy and ends in the first Riemann surface at point marked as S, where the energy is E_1 (say). It can be said that (see Eq. 6 in [10])

$$\int_{C'} d\theta_t = \int_0^{E_1} \left[\pi \frac{dn(E)}{dE} - ImTr\left(G^a \frac{\partial F^a}{\partial E} \right) \right] dE \tag{5.18}$$

We will replace C by C', when the contour is not completely closed. G^a is the advanced Greens function and F^a is self energy due to coupling the system with the leads. One can then state that $-\int_0^{E_1} ImTr\left(G^a \frac{\partial F^a}{\partial E}\right) dE$ is the correction term for Eq. (5.17) when contour C' is not closed. This statement can be alternately justified as follows. When the self energy is independent of incident energy, then the contour C is closed as well as the correction term is zero implying Eq. (5.18) becomes Eq. (5.17), where $N(E_1) = n(E_1) - n(0)$. For a double delta function potential in one dimension the correction term is very important to consider. There, energy dependence of self energy can be seen very easily in the broadening of consecutive resonance peaks (shown in Fig. 5.2d). Therefore if the approximate equality in Eq. 5.16 is to yield the exact equality in Eq. 5.17 for a closed contour C, then one can refine the statement in Eq. (5.10) to state,

$$\oint_C dE \int_{sample} d^3\mathbf{r} \frac{\delta}{\delta U(r)} = \oint_C dE \left(-\frac{d}{dE}\right) \tag{5.19}$$

where we generate a closed contour C in the Argand diagram by varying energy E from 0 to E_1. Using Eq. 5.9 on the RHS of Eq. 5.19 we get

$$\oint_C dE \int_{sample} d^3\mathbf{r} \frac{\delta}{\delta U(r)} = \oint_C dE \int_{global} d^3\mathbf{r} \frac{\delta}{\delta U(r)} \tag{5.20}$$

In case of the solid line in Fig. 5.3b, the potential everywhere is real and is 0. It traces a closed contour in the first Riemann surface. Thus for this system Eq. (5.17) is applicable. Explicit calculations of density of states for the stub show this and so everything is consistent with Eq. (5.1). However, for scattering by a delta function potential (Fig. 5.4) in a single channel quantum wire we get a counter-intuitive result. In this case, the contour of the Argand diagram (shown in Fig. 5.5a) is closed in a special way and so we expect Eq. (5.17) to be applicable. But explicit calculations of density of states in Sect. 4.2 show that is not the case, as the agreement between the dotted curve and dashed curve in Fig. 4.4 happens only at the peak and not for all E. It has been shown earlier that for this system the fundamental theorem of Büttiker-Thomas-Pretre (BTP) also breaks down [9] due to the non-analyticity of scattering matrix elements. This is a consequence of the fact that delta function potential in Q1D incorporates a log divergence in scattering matrix elements. One can see Eqs. 4.125 and 4.127 that the denominator has an infinite series that for large κ_e go as $\frac{1}{\kappa_e}$. One can also see from Eq. 4.130 that as we go on considering more and more terms in the infinite series, the energy E at which this equation is satisfied keep getting lower and lower without any bound. In case of Fig. 4.4, Eq. 4.130 is satisfied at $EW \approx 85$, that is where the curves peak corresponding to a quasi-bound state, because the infinite series was truncated at 500th term. All condensed matter systems will naturally truncate the series as very high energy transverse modes are not possible in them. Hence this is a situation where Eq. (5.1) cannot be strictly applied.

For the three prong potential, shown in Fig. 5.6, the scattering matrix is 3×3 and the correct form of FSR is given by Eq. (5.15). Whenever the scattering matrix has a dimension greater than 2, the connection between FSR in Eq. (5.15) and Eq. (5.1) is not straight forward. However, we can make this connection for each partial density of states (PDOS) and is shown below. As an example, let us consider the Argand diagrams for the three prong potential shown in Fig. 5.7a and c. None of the Argand diagrams (e.g., Fig. 5.7a, c) are closed in the first Riemann surface. Let us, for example, consider the Argand diagram for t_{31} tracing a contour C' ($PDABQFARS$ shown in Fig. 5.7c) as energy is varied from 0 to E_1. We can show that (see Appendix A)

$$\int_{C'} d\theta_{t_{31}} \approx 2\pi \int_0^{E_1} \frac{\rho_{pd}(E, 3, 1)}{|t_{31}|^2} dE \tag{5.21}$$

Now we can again state that for any closed curve like $ABQFA$ in Fig. 5.7c, Eq. (5.21) is exact. That is,

$$\oint_{ABQFA} d\theta_{t_{31}} = 2\pi \oint_{ABQFA} \frac{\rho_{pd}(E, 3, 1)}{|t_{31}|^2} dE \tag{5.22}$$

This statement can also be alternately justified as follows. For a closed contour not enclosing the singularity, L.H.S of Eq. (5.22) is zero. Also for the same closed contour, R.H.S of Eq. (5.22) will be zero as shown in Appendix B. Note that in Appendix B we have used the definition of $\rho_{pd}(E, \alpha, \beta)$ as given by R.H.S of Eq. (5.8). Hence it follows that R.H.S of Eq. (5.22) is also zero, where $\rho_{pd}(E, 3, 1)$ is given by the R.H.S of Eq. (5.8).

At present we proceed with just the equality between the integrations on LHS and RHS of Eq. 5.22 and that is enough to prove that ρ_{pd} can become negative whenever we have a Fano resonance leading to a closed sub-loop. The proof below although stated for $\rho_{pd}(E, 3, 1)$ is therefore true for any $\rho_{pd}(E, \alpha, \beta)$ inside a closed contour integral not enclosing the singularity. Now,

$$\oint_{ABQFA} d\theta_{t_{31}} = \oint_{ABQFA} \frac{d\theta_{t_{31}}}{dE} dE = 0 \tag{5.23}$$

In Fig. 5.8, $\frac{d\theta_{t_{31}}}{dE}$ is shown in the energy range covering the sub-loop $ABQFA$ of Fig. 5.7c. As is implied by Eq. (5.23), $\frac{d\theta_{t_{31}}}{dE}$ is somewhere positive and somewhere negative to ensure the area under the curve (shaded region in Fig. 5.8) is zero. The negativity of this quantity is much discussed in literature and its physical implications are inconclusive as we have discussed earlier. We can conclude from Eq. 5.22 that the R.H.S of Eq. (5.22) is zero implies that $\frac{\rho_{pd}(E,3,1)}{|t_{31}|^2}$ will also be positive as well as negative in certain energy values (or kl values). This is a conclusion that can be drawn even without explicitly calculating $\frac{\rho_{pd}(E,3,1)}{|t_{31}|^2}$, and such explicit calculations are difficult. Also $\rho_{pd}(E, 3, 1)$ as given by the R.H.S of Eq. (5.8) is conclusively negative in some energy values due to sub-loops within a Riemann surface and very unique to Q1D and such sub-loops cannot be seen in 1D, 2D, and 3D. In some cases the precise energy ranges can be identified and the negativity of the RHS of Eq. 5.8 can be further proved as will be shown in the next chapter. Although we have considered the three prong potential as an example the proof is valid for any potential whose

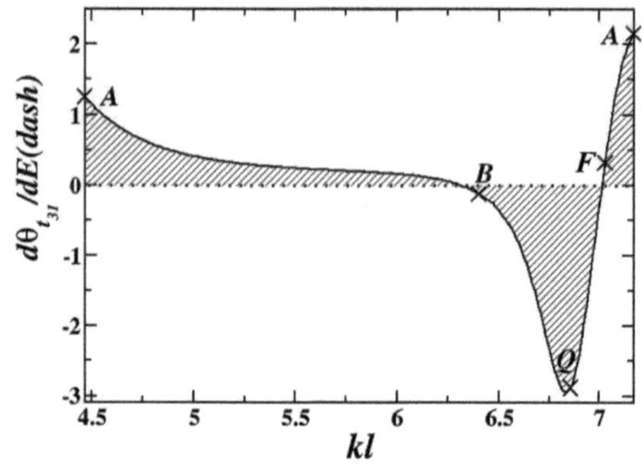

Fig. 5.8 $\frac{d\theta_{t_{31}}}{dE}$ as a function of kl for the three prong potential. Here $l = 1$, $l_2 = 5l$, $e_0 = 1$, and $e_0Ul = -1000$

Argand diagram shows a continuous phase drop due to a sub-loop as it is due to the fact that the sub-loop traces a closed contour not enclosing the phase singularity.

We will now argue how negative slopes in scattering phase shift of mesoscopic systems are fundamentally different from that studied earlier [5] in 1D. Note the negative slope at point P in Fig. 5.7d. This kind of negative slope at very low energies can arise for scattering in 1D and one can easily check this for a square well potential. In terms of our analysis we understand the negative slope at point P in Fig. 5.7d due to a convex trajectory at P in Fig. 5.7c which is originating due to the fact that the Argand diagram starts from the origin and behaves anomalously as the trajectory starting from the origin is neither clockwise nor anticlockwise with respect to the singular point (i.e., origin). See the expanded Argand diagram trajectory shown in the inset of Fig. 5.7c. The trajectory moves up, turns around, and moves down to become convex in a small energy window and then winds around the origin anticlockwise. Although $\frac{d\theta_{t_{31}}}{dE}$ is negative at P in Fig. 5.7d, there is no evidence that PDOS $\rho_{pd}(E, 3, 1)$ is negative at energy corresponding to point P (to be shown explicitly in next chapter). Such a negative slope is fundamentally different from the negative slope at Q in Fig. 5.7d that originates from a closed sub-loop $ABQFA$ in Fig. 5.7c, that we encounter only in Q1D and mesoscopic scattering. We have shown that the sub-loop seen in Fig. 5.7c implies the presence of this negative slope in scattering phase shift $\theta_{t_{31}}$ and also implies PDOS $\rho_{pd}(E, 3, 1)$ is negative. We have shown that an Argand diagram for such a scattering matrix element curls around and forms a sub-loop without violating the topological constraints of Eq. (5.1). The line integration along the sub-loop $ABQFA$ in Fig. 5.7c does not contribute to the line integration over the trajectory $PDABQFARS$ of Fig. 5.7c or I in Eq. (5.1) is unaffected by the presence or absence of a sub-loop in the closed contour C. Negative slopes of this second type (i.e., observed at Q in Fig. 5.7d) that we have discussed here in fact can appear or disappear very easily and can be found at much higher energies. In Fig. 5.9, we have shown the Argand diagram for t_{31} up to very high value of energy (or, kl). We can see many sub-loops which again implies the presence of negative slopes in scattering phase shift and negative PDOS. Sometimes there is

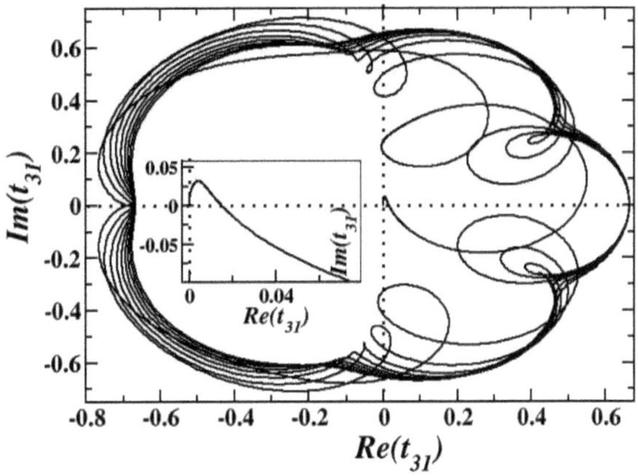

Fig. 5.9 Argand diagram for transmission amplitude t_{31}, varying kl from 0 to 50, for the three prong potential. Here $l = 1$, $l_2 = 5l$, $e_0 = 1$, and $e_0Ul = -1000$

a cusp in the Argand diagram and such a cusp means a sub-loop has disappeared. Thus the phase drop will also disappear as we go to such energies. Disappearance of a sub-loop can be demonstrated by varying any other parameter like U, l, l_2 of the three prong geometry.

Normally one would say that one should start from a closed Hamiltonian system that has well defined quantum states. The system can transmit if the incident energy matches with the energy of the state and transmission characteristics like line shapes and phase shifts are determined by the nature of the states. However, as we can see the system can have a transmission that in no way reflects the state inside the system (point P in Fig. 5.7d is a concrete example of it) but rather when it scatters then Eqs. 5.17 imply states are determined by the scattering phase shifts. Specially ρ_{pd} can never be found from the Hamiltonian alone but one has to specify from which unknown state the scattered particle came and to which unknown state it went. After all states inside the reservoir are in principle unknown. This gives one the impression that there is an anomalous regime when transmission occurs but states in the system have not developed and so open systems are more general and closed systems with well defined quantum states are a special case. This will be further elaborated in the next two chapters.

$\rho_{pd}(E, 3, 1)$ being negative is a counter-intuitive feature of quantum mechanics and can have interesting physical significance that will be discussed more in the last chapter. In plain words, it means an electron that is incident along lead 1 and transmitted to lead 3 dwells in some negative number of states (PDOS is negative) inside the scatterer. Total charge being electronic charge times number of states will be positive for these negatively charged electrons. So other electrons that are incident along lead 1 and transmitted to lead 2 or reflected back to lead 1 will be attracted by this positively behaving charge of electrons going from 1 to 3. This could be the explanation for the electron-electron attraction observed in the numerical experiment of [11], where no explanation could be given.

Having discussed negative PDOS let us come to the puzzling feature of injectance. Let us try to understand if there is a general connection between negative slope in scattering phase shift versus incident energy and semi-classical injectance becoming exact as observed for specific potential in Sect. 4.2. Let us discuss the three prong potential of Fig. 5.6 similarly as we discussed the two channel problem as done in Sect. 4.2. Thus one can understand how to generalize. Using Eq. (5.11) we can write for the three prong potential of Fig. 5.6

$$\rho_{pd}(E,3,1) \approx \frac{1}{2\pi}|t_{31}|^2\frac{d\theta_{t_{31}}}{dE} \tag{5.24}$$

$$\rho_{pd}(E,2,1) \approx \frac{1}{2\pi}|t_{21}|^2\frac{d\theta_{t_{21}}}{dE} \tag{5.25}$$

$$\rho_{pd}(E,1,1) \approx \frac{1}{2\pi}|r_{11}|^2\frac{d\theta_{r_{11}}}{dE} \tag{5.26}$$

Finally like in Eq. 4.1, we can write for the three prong potential

$$\rho(E,1) \approx \frac{1}{2\pi}|r_{11}|^2\frac{d\theta_{r_{11}}}{dE} + \frac{1}{2\pi}|t_{21}|^2\frac{d\theta_{t_{21}}}{dE} + \frac{1}{2\pi}|t_{31}|^2\frac{d\theta_{t_{31}}}{dE} \tag{5.27}$$

The R.H.S of Eq. (5.27) is the semi-classical limit of injectance. The topological interpretation of Eqs. (5.24), (5.25), (5.26) in terms of Argand diagrams leading to Eq. (5.27) is very useful. The correction term to Eq. (5.27) for three prong potential will depend only on the reflection channel (as discussed in Eq. 4.5) which gives

$$\rho(E,1) = \frac{1}{2\pi}\left[|r_{11}|^2\frac{d\theta_{r_{11}}}{dE} + |t_{21}|^2\frac{d\theta_{t_{21}}}{dE} + |t_{31}|^2\frac{d\theta_{t_{31}}}{dE} + \frac{m_0|r_{11}|}{\hbar k^2}sin(\theta_{r_{11}})\right] \tag{5.28}$$

By definition, using Eqs. 3.49 and 5.8

$$\rho^e(E,1) = \rho_{pd}(E,1,1) + \rho_{pd}(E,2,1) + \rho_{pd}(E,3,1) = \text{RHS of Eq. 5.28} \tag{5.29}$$

is exact injectance. And,

$$\rho^s(E,1) = \frac{1}{2\pi}\left[|r_{11}|^2\frac{d\theta_{r_{11}}}{dE} + |t_{21}|^2\frac{d\theta_{t_{21}}}{dE} + |t_{31}|^2\frac{d\theta_{t_{31}}}{dE}\right] \tag{5.30}$$

is semi-classical injectance (note that in Eq. 5.27 we have an approximate equality while in Eq. 5.30 an exact equality which defines ρ^s). Equation (5.28) implies that $\rho^e(E,1)$ and $\rho^s(E,1)$ will be equal at energies, where the correction term is zero or $|r_{11}|sin(\theta_{r_{11}}) = 0$. According to the earlier arguments of Leavens and Aers described at the end of Chap. 3, in the semi-classical limit $|r_{11}| \longrightarrow 0$ and $\rho^s(E,1) = \rho^e(E,1)$.

But in the case of the three prong potential, graph presented in Fig. 5.10a shows $|r_{11}| \neq 0$ at the energy corresponding to the points M and N, where one can see from Fig. 5.10b that $sin(\theta_{r_{11}})$=0 and so $\rho^s(E, 1) = \rho^e(E, 1)$. We will show how the Argand diagram topology is responsible for this behavior and therefore provides a general understanding. In Fig. 5.10a and b, we have shown the Argand diagram and phase shift for scattering matrix element r_{11}, respectively. In both the plots, kl is varied from 0 to 12. As one can see the Argand diagram in Fig. 5.10a is restricted to one side of the phase singularity (i.e., origin) and in the first Riemann surface resulting in sub-loops. This will naturally mean that the contour has both concave and convex parts in the trajectory. Scattering phase shift in Fig. 5.10b increases with kl, reaches a peak value, and then drops to become π at M. The pattern repeats as kl increases and the scattering phase shift becomes π again at N. Therefore the correction term $|r_{11}|sin(\theta_{r_{11}})$ to semi-classical injectance is zero at M and N. Thus at M and N, the semi-classical injectance (Eq. (5.30)) is exact. The exactness of semi-classical injectance is shown in Fig. 5.11 (for clarity see the inset) at points M and N corresponding to same kl values as in Fig. 5.10. For monotonously increasing phase, Argand diagram extends to higher Riemann surfaces and line integrals include the effect of other singularities in higher Riemann surface. Of course phase can be integral multiples of π (i.e., 2π, 3π, .. where ρ^s can become exact because the error from one Riemann surface cancels the error from the next Riemann surface), but for open Argand diagram trajectories as argued before, this is not a consistent behavior, except in some simple one dimensional scattering problems. This inconsistent behavior can be for example checked for an one-dimensional Aharonov-Bohm ring with different arm lengths. Drops in $\theta_{r_{11}}$ resulting in $\theta_{r_{11}}$ being π and $sin(\theta_{r_{11}}) = 0$ leading to semi-classical injectance being exact can be understood from the Argand diagram in a single Riemann surface, involving a single phase singularity in the line integral and hence is not an accident. Drops in $\theta_{r_{11}}$ is a pure quantum mechanical behavior and hence exactness of semi-classical injectance at the energies corresponding to the phase drops are counter-intuitive. As argued before, these drops coming from sub-loops in Argand diagram are tunable and can be removed by varying some parameter. In Fig. 5.12, we plot the phase behavior for the same system, after decreasing the potential U of the same system as in Figs. 5.10 and 5.11. The phase drop at point F as usual decreases to π, and hence semi-classical injectance is exact at this point which can be seen in Fig. 5.13a. Point F is marked at the same value of kl in Fig. 5.12 and 5.13a. But, in Fig. 5.12, the phase drop at G is now not sharp enough to decrease to π. It is due to tuning the potential, that the sub-loop has now reduced in area, and consequently the drop has also reduced and will eventually be reduced to a cusp as U is decreased further. At such point $\rho^e(E, 1)$ is not equal to $\rho^s(E, 1)$, i.e., semi-classical injectance is not exact in spite of a phase drop. This can be seen in Fig. 5.13b where point G is marked at the same value of kl, as in Fig. 5.12. One would have expected that when we decrease the potential and make it weaker, semi-classical behavior will be favored. But on the contrary, for stronger potential the semi-classical injectance is exact at point N in Fig. 5.10b, while for a weaker potential in the same system the semi-classical injectance is not exact at point G in Fig. 5.12. Therefore, the drops in scattering phase shift of mesoscopic system, originating from

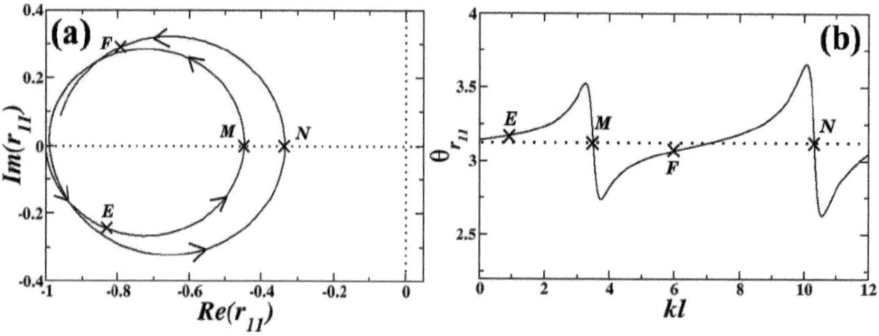

Fig. 5.10 **a** Argand diagram for reflection amplitude r_{11} and **b** plot of reflection phase shift $\theta_{r_{11}}$, versus kl for the three prong potential. Here $l = 1, l_2 = 5l, e_0 = 1$, and $e_0Ul = -10000$

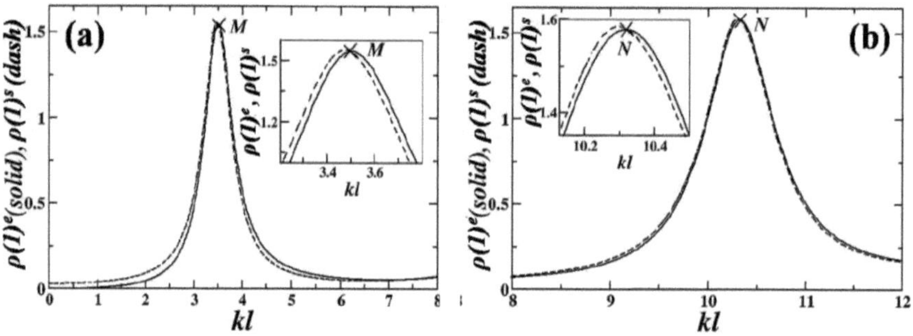

Fig. 5.11 Plot of exact injectance $\rho^e(E, 1)$ (solid line) and semi-classical injectance $\rho^s(E, 1)$ (dashed line) as a function of kl for the three prong potential. The peaks in the injectance are shown separately, **a** shows the first peak, **b** shows the second peak, for the same parameters as in Fig. 5.10. The insets show the magnified curves at points M and N

sub-loops in Argand diagram, involve completely new physics. One has to discard the usual concept of semi-classical regimes wherein the de-Broglie wavelength of the electron is much smaller than the scale of the potential, mathematically expressed by Eq. (5.10).

5.4 Some Comments on Experiments

Argand diagram of scattering matrix elements can also be done for experimental data. We look at the data published in Fig. 3 by Y. Ji et al. [Science, 2000] and with a line gauge recover the numerical values of the data and plot them in the form of an AD in Fig. 5.14 [6]. In the order of Fig. 5.14D, C, B, and A, we can see a disappearing sub-loop in one Riemann surface. The sub-loop is the largest in Fig. 5.14D and completely confined to the upper half plane. All sub-loops avoid the origin. So in this regime

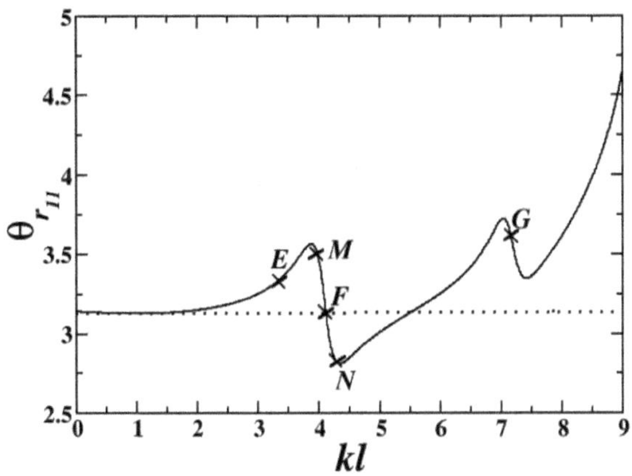

Fig. 5.12 Plot of reflection phase shift $\theta_{r_{11}}$ versus kl for the three prong potential. Here $l = 1$, $l_2 = 5l$, $e_0 = 1$, and $e_0 U l = -1000$

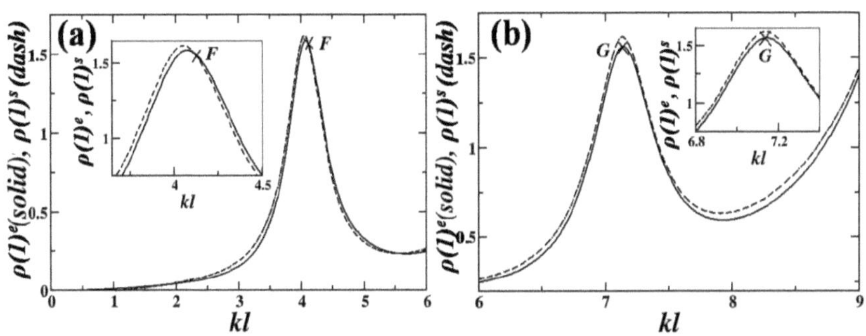

Fig. 5.13 Plot of exact injectance $\rho^e(E, 1)$ (solid line) and semi-classical injectance $\rho^s(E, 1)$ (dashed line) as a function of kl for the three prong potential. The peaks in the injectance are shown separately, **a** shows the first peak, **b** shows the second peak, for the same parameters as in Fig. 5.12. The insets show the magnified curves at points F and G

one may design experiments to test other aspects related to DOS hierarchy. Some such aspects will be discussed in the next two chapters. Similar exercise on Fig. 3 of the publication by Zaffalon et al. [Phys. Rev. Lett. 2008] gives the AD shown in Fig. 5.15. According to Zaffalon et al. the system is in the Kondo regime. Notice that in Figs. 5.2b and 5.7a we need two resonances for the AD to cover one Riemann surface. Hence the AD shows two lobes around the origin. In the Kondo regime, spin splitting will make this four lobes in the first Riemann surface around the origin as schematically shown in Fig. 5.16. Thus the AD, BC, analyticity, and the fact that states are essentially created by a winding number give us an intuitive understanding of how scattering phase shift will behave for a Kondo resonance. Accordingly, we can conclude some qualitative features of DOS hierarchy for a Kondo resonance and thus get some idea about its thermodynamic properties.

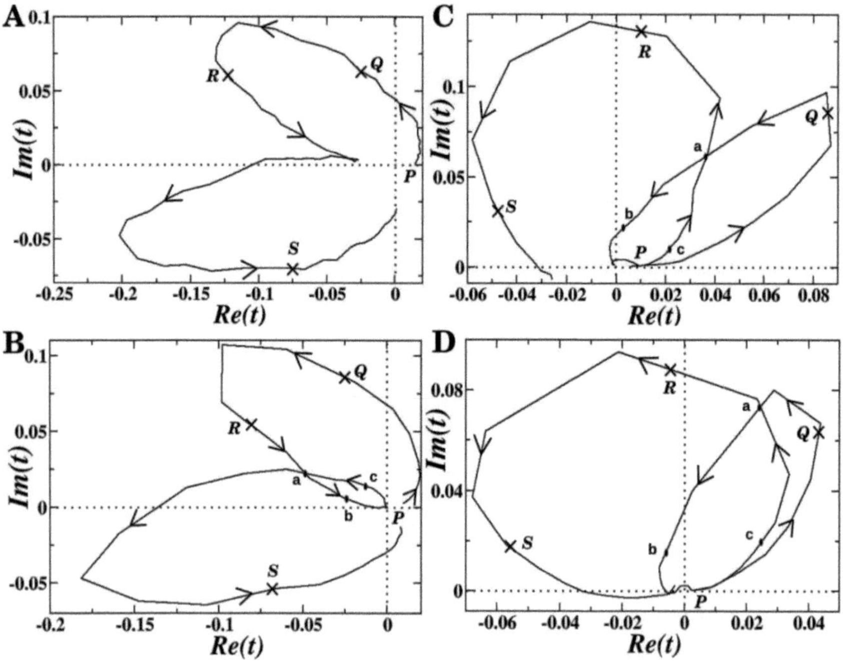

Fig. 5.14 We re-plot the data in Fig. 3 of Ji et al. [Science 2000] in the form of AD

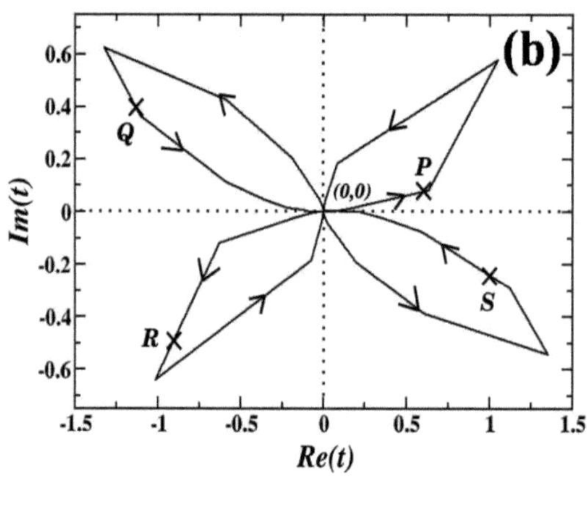

Fig. 5.15 We re-plot the data in Fig. 3 of Zaffalon et al. [Phys. Rev. Lett. 2008] in the form of AD

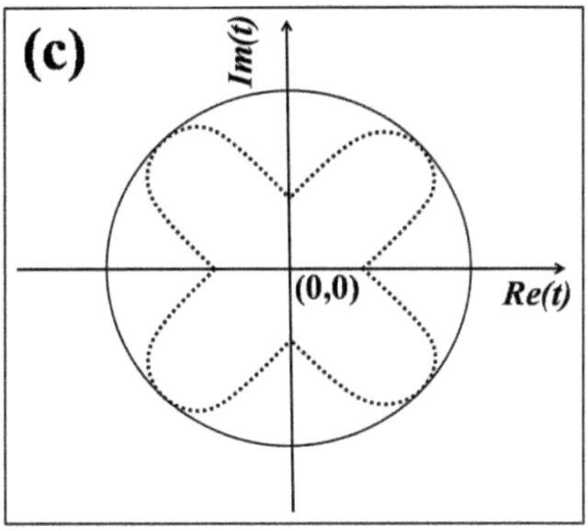

Fig. 5.16 An intuitive understanding of AD for Kondo resonances

5.5 Quantum Capacitance

Equation 5.22 can be further written as

$$\oint_{ABQFA} d\theta_{t_{31}} = 2\pi \oint_{ABQFA} \frac{\int_{sample} d^3\mathbf{r}\rho_{lpd}(E, 3, \mathbf{r}, 1)}{|t_{31}|^2} dE \tag{5.31}$$

or from Eq. 3.45

$$\oint_{ABQFA} d\theta_{t_{31}} = -\oint_{ABQFA} dE \int_{sample} d^3\mathbf{r}\frac{\delta\theta_{t_{31}}}{\delta U(\mathbf{r})} \tag{5.32}$$

Note that the closed contour is generated by varying E and the equality holds for the contour integral. However we will see in the next chapter that there are situations when the closed contour or the sub-loop is such that it is smoothly coming back to the same point in the complex plane within the first Riemann surface. In such cases that we will refer to as smooth sub-loops, even the integrands become equal. For a general situation, suppose to a good approximation we can write

$$\int_{A\to B} d\theta_{t_{31}} \approx -\int_{A\to B} dE \int_{sample} d^3\mathbf{r}\frac{\delta\theta_{t_{31}}}{\delta U(\mathbf{r})} \tag{5.33}$$

Since the points A and B are close as $A \to B$, we can write

$$\frac{d\theta_{t_{31}}}{dE} \approx -\int_{sample} d^3\mathbf{r}\frac{\delta\theta_{t_{31}}}{\delta U(\mathbf{r})} \tag{5.34}$$

which is also stated in Eq. 3.54 but now we are saying that this is a good approximation when energy range is within a closed sub-loop, or in other words near a Fano resonance. In fact there can be some resonances for which it is exact and that opens up a new avenue for exploring many body effects as shown below. Equation 5.34 implies

$$\frac{d\theta_{t31}}{dE} \approx -\frac{\delta\theta_{t31}}{\delta U_{g_1}} - \frac{\delta\theta_{t31}}{\delta U_{g_2}} - \cdots \tag{5.35}$$

where U_{g_n} are a spatial series of gate voltages in small regions of the sample. Thus a functional derivative is behaving like partial derivative in the sense that an exact derivative can be written as a sum of functional derivatives as if the local potentials in different points of the sample are behaving like independent variables of the theory. Specially because the LHS of Eq. 5.35 can be compared to a constant shift of potential within the sample region and we know that around a Fano resonance this becomes exact. This has to do with the fact that there is a unique amount of time that the electron spends under each gate voltage and they can be added up to get the total time the particle takes to traverse the sample. Hence this is a consequence of the property of time although mathematically functional derivative and partial derivative are different if we are not considering scattering phase shifts in particular. This is valid even in presence of electron-electron interaction in the sample with no interaction in the leads, and a numerical verification of it can be seen in [12].

Suppose there are M electrons at coordinates $r_1, r_2, \cdots, r_M$ inside a closed system. Let an STM tip make a small perturbation $\sum_n \Delta U \delta(r - r_n)$ at the point r inside the sample. Then we can write an expansion

$$\Delta E = \sum_n \int d^3r_1 d^3r_2 ... d^3r_M \psi^*(r_1, r_2, \cdots, r_M) \Delta U \delta(r - r_n) \psi(r_1, r_2, \cdots, r_M) + O(\Delta U^2) + O(\Delta U^3) + \cdots \tag{5.36}$$

$$\text{as } \Delta U \to 0 \text{ then } \frac{\partial E}{\partial U(r)} = Q(r) \text{ for a fixed point } r \tag{5.37}$$

$$\text{where } Q(r) = M \int d^3r_2 d^3r_3 ... d^3r_M \psi^*(r, r_2, ...r_M) \psi(r, r_2, ...r_M) \tag{5.38}$$

$\psi^*(r, r_2, ...r_M)$ being the many body wave-function. We can define a Lindhard function $\eta(r)$

$$\eta(r) = -\frac{\partial Q(r)}{\partial U(r)} \tag{5.39}$$

and $Q(r)$ is the same as the LDOS at the point r at zero temperature. From Eq. 5.37

$$\eta(r) = -\frac{\partial^2 E}{\partial U(r)^2} \tag{5.40}$$

We can polarize a closed system by applying such a local potential at a fixed point r, as well as by applying an AB flux ϕ. Now that will create a change of electrostatic potential $\delta U(r')$ at all points in the sample and so the many body effects will self consistently define this change which will be infinitesimal because the applied potential at the point r is infinitesimal. As we have seen that functional derivatives are similar to partial derivatives as if the changes of potential at different points are like the independent variables of the system, we can write Eq. 5.41 below. While this was shown in absence of interactions in Eq. 5.35, now we are claiming that in presence of the self consistently defined $\delta U(r')$ and this was numerically verified by [12]. Since these statements are for infinitesimal changes, they are also intuitively expected if we are not close to a phase transition and in such small systems we need not worry about that. This also shows that the idea of a local time that adds up to give a total time is applicable in presence of many body effects.

$$dQ(r) = \frac{\partial Q(r)}{\partial \phi} d\phi + \frac{\partial Q(r)}{\partial U(r)} dU(r) + \int_{r' \neq r} \frac{\partial Q(r)}{\partial U(r')} dU(r') d^3 r' \tag{5.41}$$

As stated before we can replace the last term by a functional derivative

$$dQ(r) = \frac{\partial Q(r)}{\partial \phi} d\phi + \frac{\partial Q(r)}{\partial U(r)} dU(r) + \int_{r' \neq r} \frac{\delta Q(r)}{\delta U(r')} \delta U(r') d^3 r' \tag{5.42}$$

In discrete form this can be written as

$$dQ_i = \frac{\partial Q_i}{\partial \phi} d\phi + \frac{\partial Q_i}{\partial U_i} dU_i + \sum_{j \neq i} \frac{\delta Q_i}{\delta U_j} \delta U_j \tag{5.43}$$

To a linear order we can write

$$dQ_i = C\,(dU_i - dU_{rest}) \tag{5.44}$$

where C is just the geometric capacitance. This means the polarization creates an effective potential difference $(dU_i - dU_{rest})$ between the test site i and the rest of the system. That means the functional change δU_j at different sites j in the system can be replaced by an average potential dU_{rest}. From charge conservation we can write

$$-\eta = \frac{\partial Q_i}{\partial U_i} = -\sum_{j \neq i} \frac{\delta Q_i}{\delta U_j} \tag{5.45}$$

This is because we can apply a voltage U_i at the point i to deplete the charge there, and this displaced charge will go to the rest of the system, the distribution of which will be self consistently determined by many body effects. That will create the self consistent potential δU_j at all points j. In other words we may apply a suitable potential δU_j at all points j to determine the same polarization charge at the point i. Thus from Eqs. 5.43, 5.44, and 5.45

$$C\left(dU_i - dU_{rest}\right) = \frac{\partial Q_i}{\partial \phi} d\phi - \eta dU_i + \eta dU_{rest} \tag{5.46}$$

$$(C+\eta)\frac{dU_i - dU_{rest}}{d\phi} = \frac{\partial Q_i}{\partial \phi} = \frac{\partial Q_i}{\partial U_{rest}} \frac{\partial U_{rest}}{\partial \phi} \tag{5.47}$$

Now the polarization is created by U_i and flux ϕ and so we can write U_{rest} as a function of these two independent variables. That is

$$dU_{rest} = \frac{\partial U_{rest}}{\partial \phi} d\phi + \frac{\partial U_{rest}}{\partial U_i} dU_i \tag{5.48}$$

Now since the site i is an isolated point, one may write $\frac{\partial U_{rest}}{\partial U_i} \rightarrow 0$. Thus from Eq. 5.47 we get

$$(C+\eta)\frac{dU_i - dU_{rest}}{d\phi} = \eta \frac{\partial U_{rest}}{\partial \phi} = \eta \frac{dU_{rest}}{d\phi} \tag{5.49}$$

Multiplying both sides by C we get

$$C\frac{dU_i - dU_{rest}}{d\phi} = \frac{C\eta}{C+\eta} \frac{dU_{rest}}{d\phi} \tag{5.50}$$

$$C(dU_i - dU_{rest}) = \frac{C\eta}{C+\eta} dU_{rest} \tag{5.51}$$

$$dQ_i = C_{eff} dU_{rest} \tag{5.52}$$

$$\frac{1}{C_{eff}} = \frac{1}{C} + \frac{1}{\eta} \tag{5.53}$$

Hence there is a quantum capacitance η in series with the classical capacitance C. This result was known for non-interacting electrons but it is meaningless to talk of capacitance in absence of electron-electron repulsion. The notion of time in quantum mechanics tells us that close to Fano resonances or within smooth closed sub-loops where functional derivative is similar to a partial derivative, this result will be valid in presence of electron-electron repulsion. Away from the Fano resonance again this principle enforced by the nature of time should again be applicable but to see a systematic behavior we need to extend the analysis beyond a single Riemann surface and how multiple phase singularities contribute to the winding number.

Given that $\int d^3\mathbf{r}\frac{\delta\theta_{t31}}{\delta U(\mathbf{r})}$ is a winding number, as can be seen from Eq. 5.32, if we can generate a smooth sub-loop by varying the potential $U(r)$ inside the sample only, then Eq. 5.35 will be exact. Since winding number is a conserved quantity, Eq. 5.35 tells us that we may generate this winding number by varying E, or by varying $U(r)$, or by varying just one of the gate voltages, say U_i. Although the contours will be different, they all give the same winding number or the same number of local states associated with a particular Fano resonance with a particular number of nodes. That is winding number does not care if we generate the closed loop by varying just one of the U_{g_n} or we vary many of them. Thus Eq. 5.32 imply that if we plot $\theta_{t_{31}}$ as a function of energy E then it will give us the winding number $\int d^3\mathbf{r}\frac{\delta\theta_{t31}}{\delta U(\mathbf{r})}$ and the integration over $\mathbf{r}$ can be over a small region of the sample. Therefore, in Fig. 7.4 too the variation in $\theta_{t_{31}}$ is determined by (or determines) this winding number, and it relates to an experimentally observable that we will discuss in Chap. 7 in detail.

5.6 Summary

Several conclusions can be drawn from the Argand diagrams without referring to the Hamiltonian or to the scattering potential. In 1D, 2D, and 3D, the Argand diagram trajectory encircles the phase singularity. But in mesoscopic systems we find that the Argand diagram can develop sub-loops. A sub-loop does not enclose the phase singularity at the origin and hence topologically allowed. A sub-loop can therefore appear or disappear on varying some parameter. It does not matter what parameter (say E) is varied to obtain the Argand diagram or what parameter (say U) is varied to make the sub-loop disappear. When the sub-loop appears there will be a gradual drop in the scattering phase shift and when the sub-loop disappears the drop will also disappear. Hence appearance and disappearance of phase drop are also very natural and pose no conceptual problem. Just as the sub-loop appears on varying some parameter, it can also grow in size as the parameter is varied. As the sub-loop becomes large and comes closer to the origin the phase drops also become large and sharp making the scattering phase shift decrease to π or zero (phase shift is reset to initial value independent of the initial phase of the incident electron wave-function) and then semi-classical injectance or Friedel sum rule becomes exact. This is very counter-intuitive as the strong phase drop signifies onset of pure quantum mechanical behavior while the Friedel sum rule is expected to become exact in semi-classical regimes.

Also we prove that whenever there is a sub-loop (big or small) there will be negative partial density of states. For example if there is a sub-loop in the range ΔE then there is also negative partial density of states in the range ΔE. Conclusive evidence of negative partial density of states in a real system has never been reported before. Since all these results are drawn from the properties of the Argand diagram, the results are general and independent of the Hamiltonian or the scattering potential. The physics originating from sub-loops is completely new and upsets our way of understanding semi-classical behavior.

The fact that PDOS can be determined exactly from semi-classical formula at the Fano resonances can be very useful. Because an experimentalist can just vary the Fermi energy of the incident particle in a typical mesoscopic scattering scenario and measure the scattering phase shift as was demonstrated by [[4]; Kobayashi et al. (2004)]. From the energy derivative of the phase shift the PDOS can be determined without any knowledge of the exact potential profile and impurity configuration inside the system. Thus several members in the hierarchy of DOS that determine thermodynamic and transport properties of the sample can be known. This will be specially useful because mesoscopic systems are not in the ergodic regime allowing us to average over impurity configurations.

5.7 Appendix A

From Eq. (5.8), we can write

$$\rho_{pd}(3,1) = -\frac{1}{2\pi}\int_{sample} d^3\mathbf{r}|t_{31}|^2\frac{\delta\theta_{t_{31}}}{\delta U(\mathbf{r})} \tag{5.54}$$

In Eq. (5.54) using the well known semi-classical approximation $\int_{sample} d^3\mathbf{r}\frac{\delta\theta_{t_{31}}}{\delta U(\mathbf{r})} \cong -\frac{d\theta_{t_{31}}}{dE}$, we get (see Eq. 5.11)

$$\rho_{pd}(3,1) \approx \frac{1}{2\pi}|t_{31}|^2\frac{d\theta_{t_{31}}}{dE} \tag{5.55}$$

Now, for a contour C' traced when energy is varied from 0 to E_1,

$$\int_{C'} d\theta_{t_{31}} = \int_0^{E_1}\frac{d\theta_{t_{31}}}{dE}dE = \int_0^{E_1}\frac{\frac{1}{2\pi}|t_{31}|^2\frac{d\theta_{t_{31}}}{dE}}{\frac{1}{2\pi}|t_{31}|^2}dE$$

Using Eq. (5.55)

$$\int_{C'} d\theta_{t_{31}} \approx 2\pi\int_0^{E_1}\frac{\rho_{pd}(3,1)}{|t_{31}|^2}dE \tag{5.56}$$

5.8 Appendix B

We know that in a scattering problem increasing incident energy by dE is equivalent to decreasing the potential globally by a constant amount $\Delta\varepsilon$, such that $dE = -e_0\Delta\varepsilon$, where e_0 is particle charge that we will set to 1 to simplify our arguments. That is, the new potential is $U'(\mathbf{r}) = U(\mathbf{r}) - \Delta\varepsilon$. Hence if we can generate a closed sub-loop in the Argand diagram by varying E, then we can also do so by globally changing the potential and for such a closed contour like $ABQFA$ in Fig. 5.7c,

$$\oint\limits_{ABQFA} \delta\theta_{s_{\alpha\beta}} = 0$$

$$\text{or} \quad \oint\limits_{ABQFA} \delta\theta_{s_{\alpha\beta}} = -\oint\limits_{ABQFA}\int\limits_{global} \frac{\delta\theta_{s_{\alpha\beta}}}{\delta U(\mathbf{r})}\Delta\varepsilon d^3\mathbf{r} = 0 \tag{5.57}$$

Now we replace the global integration over $\mathbf{r}$ by an integration over the sample or the scattering region only, since we have seen that it can be done in case of closed contours or inside an integration of the type $\oint\limits_{C}$ in Eq. (5.57). This has been discussed in Eq. (5.20).

$$\text{or} \quad \oint\limits_{ABQFA} \delta\theta_{s_{\alpha\beta}} = -\oint\limits_{C}\int\limits_{sample} \frac{|s_{\alpha\beta}|^2}{|s_{\alpha\beta}|^2}\frac{\delta\theta_{s_{\alpha\beta}}}{\delta U(\mathbf{r})}\Delta\varepsilon d^3\mathbf{r} = 0$$

$$\text{or} \quad \oint\limits_{ABQFA} \delta\theta_{s_{\alpha\beta}} = -\oint\limits_{C}\frac{1}{|s_{\alpha\beta}|^2}\Delta\varepsilon \int\limits_{sample} |s_{\alpha\beta}|^2\frac{\delta\theta_{s_{\alpha\beta}}}{\delta U(\mathbf{r})} d^3\mathbf{r} = 0$$

$$\text{or} \quad \oint\limits_{ABQFA} \delta\theta_{s_{\alpha\beta}} = 2\pi\oint\limits_{C}\frac{\rho_{pd}(\alpha,\beta)}{|s_{\alpha\beta}|^2}\Delta\varepsilon = 0 \quad \text{Using Eq. (5.8)}$$

In the last step we are identifying ρ_{pd} of Chap. 3, for one electron, with the I of Eq. 5.1, while ρ_{pd} is ρ_{lpd} integrated over the coordinates of the sample only. Therefore, the R.H.S of Eq. (5.22) is justified if an electronic charge is multiplied to the numerator.

References

1. Nye JF, Berry MV (1974) Proc R Soc Lond A 336:165–190; Berry MV (1998) J Mod Opt 45:1845–1858
2. Englman R, Yahalom A (2000) Phys Rev B 61:2716
3. Jayannavar AM (1994) Phys Rev B 49:14718

4. Kobayashi K, Aikawa H, Katsumoto S, Iye Y (2003) Phys Rev B 68:235304; Kobayashi K, Aikawa H, Sano A, Katsumoto S, Iye Y (2004) Phys Rev B 70:035319
5. Landauer R, Martin Th (1994) Rev Mod Phys 66:217–228
6. Satpathi U, Singha Deo P (2016) Ann Phys 375:491
7. Schuster R, Buks E, Heiblum M, Mahalu D, Umansky V, Shtrikman H (1997) Nature (London) 385:417–420
8. Singha Deo P (1998) Solid State Commun 107:69
9. Singha Deo P (2005) Int J Mod Phys B 19:899
10. Yeyati AL, Buttiker M (2000) Phys Rev B 62:7307
11. Singha Deo P (2002) Pramana J of Phys 58:195
12. Mukherjee S et al (2011) Physica E 44:62
13. Ji Y, Heiblum M, Sprinzak D, Mahalu D, Shtrikman H (2000) Science 290:779
14. Zafflon M, Bid A, Heiblum M, Mahalu D, Umansky V (2008) Phys Rev Lett 100:226601

Chapter 6
Time Travel

Abstract So far we have established that although PDOS is not defined in conventional quantum mechanics, it is well defined in functional analysis as a winding number of the AD. For a closed loop the winding number is an integer while for a part of the loop the winding number is a fraction that can be represented by a real number. It becomes clear that PDOS as a winding number can be also negative. This definition coming from functional analysis is thereby mathematical and does not rely on the questionable physical idea of spin precession. We will show in this chapter that negative PDOS will also lead to transmitting a physical signal in negative time. Once again the detail of the proof depends on BC and is independent of the potential or model. The only requirement is that the system should sustain Fano resonances and around such a resonance the stationary phase approximation is also analyzed to show that time travel is natural. We will comment on different notions of time and relate them to signal propagation time. We end the chapter with a small section called "reflections" which intends to imply that we are aware of the philosophical consequences but we are not willing to engage in a philosophical discussion at the moment. But we have hinted on what would be a physicists approach to such a philosophy. In the next chapter we will provide much more practical suggestions on how negative PDOS can manifest in the laboratory.

6.1 Different Scenarios of Time

In the Newtonian picture, time is a parameter that changes by a constant magnitude (no reference of a direction) $\Delta\tau^N$, at every point in our 3D universe to give the universe at next time or next instant. To do full justice to the idea of a next in a continuum one needs an elaborate course on differential geometry and functional analysis and Newtonian calculus often lacks the necessary sophistication. Quantum mechanics on the other hand uses this idea of time as a parameter in 3D space, but does a good job of treating the continuum via von-Neumann approach. In relativity the only time that makes sense is time interval clocked by local clocks in which a signal is sent, and every point in space has its own clock. That being a complete hotch-potch experimentalists doing experiments on suspected superluminality in quantum

P. S. Deo and K. Meena, *Time Travel: A Reality in Mesoscopic Physics*,
https://doi.org/10.1007/978-981-96-1136-2_6

mechanics are often dabbling in expressions to describe their work. In the Newtonian picture if a bullet comes at someone faster than the speed of light then one may say that the bullet cannot be seen approaching (even by an appropriate detector) as it has overtaken its own light, but the bullet will hit the person anyway because the time at the coordinate of the person in 3D world is changing identically to the time at the instantaneous coordinate of the bullet. Whereas if a real experiment is done, then we know from STR that the bullet has no chance of hitting the person as it goes outside the light cone. The gun may be inside the light cone and the target may also be inside the light cone. But as the bullet accelerates due to gunpowder inside the light cone, will go out of the light cone. Of course at some point the bullet can decelerate and get back inside the light cone and if it really wants to hit the target then these accelerations and decelerations have to be accurately taken into account. The superluminality experiments routinely claim that their classical detector detects a superluminal electro-magnetic signal coming from a classical source [1]. Time in quantum mechanics appear as a parameter and there is no self-adjoint time operator consistently defined yet. This is not a serious problem as experimentalists only measure time intervals. This suits mesoscopic physics wherein we have a system which acts as a quantum scatterer coupled to reservoirs via perfectly conducting leads. The reservoirs are completely thermalized and classical, hence inside the reservoirs time will be just the classically recorded time given by the phases of the sun and the moon tabulated by calenders and clocks. The Landauer picture goes hand in hand with von-Neumann quantum mechanics that says an experimentalist will generate a pulse of electrons which in classical world is a spray or a shot or a pulse. The source reservoir being classical has to generate this pulse and throw it into the connected lead. Inside the lead wave mechanics takes over and the pulse has to be represented as a wave-packet. The reservoirs also ensure that states do not naturally give rise to cross terms in the leads and this results in textbook style scattering phenomenon that can be studied by constructing an incident wave-packet or alternately by partial waves (Copenhagen interpretation). The electron propagation from one lead to another lead takes place via scattering through the sample and this dynamics is purely guided by quantum mechanics. Since there is no return from the sink reservoir, a propagation time from source reservoir to sink reservoir is the same as a traversal time from one reservoir to another and naturally does not violate any principle. It is to be noted that this sharp demarcation between the classical world and the quantum world is a model that solves the quantum measurement problem in mesoscopic systems and it has experimental validation. While Schrodinger cat paradox suggests that measurement results can probabilistically reveal a dead cat or a live cat, a measurement in mesoscopic systems reproducibly reveals a two probe conductance or a three probe conductance or a four probe conductance. At the same time for mesoscopic systems there is no objective conductance as one would expect from classical physics wherein a cat is either objectively dead or objectively alive. There is either a two probe conductance or a three probe conductance or a four probe conductance and what we measure depends on how we measure. However, other problems remain and that is what we intend to address in this chapter. Quantum mechanics give this measured time intervals very correctly in the semi-classical limit [2]. The low energy quantum limit is so far not understood in 1D, 2D, and 3D since calculated time interval is not always consistent with the Copenhagen interpretation of quantum mechanics which demand that the time evolution of a wave-packet can be analyzed through partial waves. There are even experiments that confirm the fact that a time-dependent wave-packet of a few particles, scattered repeatedly, produces the same interference pattern as a monochromatic stationary partial wave. It is known that quantum mechanics starts from a very different set of axioms and does not have to respect theory of relativity within the single particle coherence length. At the low energies of the mesoscopic systems, Schroedinger equation determines the dynamics within this single particle coherence length and it is completely independent of the speed of light. Beyond this coherence length all phenomenon have to respect the theory of relativity and information speed is limited by the speed of light. We go beyond 1D, 2D, and 3D and look into quasi-1D (Q1D).

6.2 Traversal Time is Propagation Time in Mesoscopic Physics

There are two traversal times defined in quantum mechanics that are called Larmor precession time (LPT) and Wigner delay time (WDT). There are many other ways of defining time that correspond to different physical situations, but they cannot be called traversal time because they are not consistent with the number of intermediate states through which a particle is propagated or traversed in time. Our definition of traversal time is that it has to be consistent with the density of states (DOS) up to a constant factor of h (Plank's constant). There was a confusion whether in quantum mechanics, the traversal time gives the first passage time or the propagation time. According to the previous section since there is no return for a particle from the sink reservoir in a mesoscopic set up, this traversal time will also be the wave-packet propagation time from source reservoir to the sink reservoir and that will be the subject of discussion in this chapter. This definition is consistent with the definition of traversal time for tunneling under the barrier [3] where one considers the time interval in which the tunneling particle interacts with the entire length of the barrier. For tunneling modes higher order multiple reflections or higher order Feynman paths are more strongly damped and so only the length of the barrier matters. We are not considering the situation of tunneling but resonant situations, namely Fano resonances. Our subject of study is the time interval consistent with DOS and how they are related to the propagation of a wave-packet. Thus the only two candidates are LPT and WDT. The two candidates for traversal time may have different mathematical forms but should lead to the same quantitative value in order to be consistent. The reason for their different forms is explained below.

6.3 Known Facts in 1D, 2D, and 3D

In this section, we restrict our discussion to 1D for simplicity, but the facts are also true in 2D or 3D. The derivation of LPT in Chap. 3 starts from a monochromatic plane wave or a stationary state and derives an expression for traversal time. Hence it starts from the time independent Schrodinger equation and derives a time wherein the parameter for time in the Schrodinger equation do not enter the analysis in any way. According to Chap. 3, LPT is physically the average traversal time of electrons in the stationary beam and such electrons in a stationary beam cannot be used to send a signal. However when summed up over the indices and coordinates, it is completely consistent with DOS at all energies. Question arises if LPT, without the summing, is at all a physical object or even a mathematical object apart from being a first-order derivative of scattering matrix element. The whole idea of spin precession can be questioned. One can always define a second derivative or a third derivative or an arbitrary order of derivative of scattering matrix element but do they mean anything beyond just a derivative that one can calculate. For example, consider the fact that first-order derivative to a manifold in differential geometry defines an object called a tangent space. Similarly, the inverse of the double derivative of a stock market curve quantifies curvature of something changing. These are examples wherein complex mathematical objects emerge from simpler ones. In Chap. 5 we can assign an objective meaning to LPT only when we find closed sub-loops in AD within a single Riemann surface and it turns out to be the winding number. For a full closed sub-loop the winding number or LPT is a natural number while for a part sub-loop it is a fraction that can be quantified by some real number, that could be positive as well as negative. Well defined mathematical objects have an uncanny habit of manifesting physically. In this chapter we will relate LPT to signal propagation time and as a consequence even a signal propagation in negative time. In the next chapter we will show other much simpler physical manifestations of its negativity. In case of a complete loop that extends beyond one Riemann surface, one may still extend this objectification but that is an open question. One has to essentially figure out how to separate out the contributions coming from the different Riemann surfaces. In such cases LPT may still be physical but we will show that the kind of experiments we discuss in the next chapter cannot bring that to light.

If not restricted by having to carry a signal or information, even in classical physics one can exceed the speed of light. A signal can be sent by using a wave-packet for example a large or a small wave-packet can mean different things. Derivation of WDT actually considers the propagation of a wave-packet in time to derive a traversal time subject to the fact that the wave-packet satisfies stationary phase approximation. Only when this approximation is valid, the parameter of time in Schrodinger equation can be related to propagation time of wave-packet or the signal propagation time. If stationary phase approximation is valid at some energy range then the LPT summed over coordinates and the WDT are quantitatively the same, positive definite, and fully consistent with DOS. In this regime, therefore, a signal can be propagated in LPT as it is identical to WDT. This happens at energies when quantum effects are moderate confirming the Copenhagen interpretation wherein a scattering problem can be seen as scattering of a wave-packet or alternately as a scattering of stationary partial waves. At low energies when quantum effects dominate, then LPT and WDT do not give the same result (as will be shown in Fig. 6.5) and in this regime they sometimes become negative, though differently (see Fig. 6.5). WDT is approximate and can become negative but since at these low energies WDT deviates from LPT, it is no longer consistent with DOS and so its negativity need not mean anything physical, specially as stationary phase approximation is involved. We will show that in Q1D there are wide and robust energy ranges wherein LPT and WDT can be both negative, identical to each other, and hence both are fully consistent with DOS. One can always claim that is a confirmation of the fact that one can send a signal in LPT and therefore in negative time. The reason for that is the basic fact that it does not make sense to solve the time-dependent Schrodinger (TDS) equation at all to study the propagation of a wave-packet. First because the time that appears in the TDS equation cannot be called physical time but rather a parameter in the sense of Newtonian mechanics. And secondly it can only be solved numerically (except for ideal cases) and the numerical algorithms like the Gauss-Siedel method or fast Fourier transformation method come up with very algorithm specific way to handle this parameter. These algorithms make sense only in the limit of time tending to infinity. Scattering phase shifts cannot even be defined directly from the time-dependent Schrodinger equation and hence one decomposes the wave-packet into partial waves that satisfy the time-independent Schrodinger equation. From the solutions of time-independent Schrodinger equation, the wave-packet can be reconstructed at some later time. We will, however, also argue consistency with stationary phase approximation, and hence their negativity does correspond to sending a signal in negative time. In the next chapter we will show that negative LPT has much simpler experimental manifestations through LPDOS.

We describe below the derivation of WDT as we want to clearly outline the nature of the stationary phase approximation used. Consider a one-dimensional wave-packet at a spatial coordinate x and a temporal coordinate τ that are with respect to a suitable choice of origin of the respective coordinates.

$$u(x,\tau) = \frac{1}{\sqrt{2\pi}} \int_{-\infty}^{\infty} a(k) e^{i(kx-\omega\tau)} dk \tag{6.1}$$

Here $a(k)$ is the weight of the kth component $e^{i(kx-\omega\tau)}$. As mentioned in the introduction of Chap. 1, in the von-Neumann picture, to start with, this wave-packet has nothing to do with solutions of Schrodinger equation and can represent a pulse of electrons or a pulse of classical electro-magnetic waves coming from source (which in mesoscopic physics is the classical source reservoir), with equal accuracy. But the pulses will evolve differently in the two cases, the time evolution of classical electro-magnetic waves being determined by Maxwell's equations, that are always relativistic and have no non-relativistic limit. As the wave-packet evolves, the τ for classical electromagnetic waves is the precise time defined by GTR, while the τ in quantum mechanics is just a parameter very similar to the Newtonian world view. There is always a subtle difference wherein the x for an electromagnetic pulse represents a physical displacement while for an electron pulse $e^{i(kx-\omega\tau)}$ is just an element of a special function space called Hilbert space, and we do not have to be very conscious of this difference at a practical level because this difference is only at the level of representations. The von-Neumann picture answers the question, what should be the differential equation that will represent the time evolution of the electron wave-packet in the leads and sample till it reaches the classical sink reservoir at the other end. The question can be answered (at least in the present case where there is no time-dependent term in the Hamiltonian) in terms of the components of the wave-packet that is found to satisfy the time-independent Schrodinger equation from which the wave-packet can be constructed at any intermediate point and thus we get a time evolution of the wave-packet. So the von-Neumann philosophy can be expressed in a nut shell by saying that mother nature will generate a pulse through an experimentalist, and theoreticians will try to predict the time evolution of the pulse for which they find a differential equation called the time-dependent Schrodinger equation. That Schrodinger equation also give a dispersion relation between k and ω that is different from that of electromagnetic waves. If the wave-packet disperses too much then it cannot carry a signal and so signal propagation time may or may not be different from the simple time evolution of the wave-packet. In GTR, or for a propagating electromagnetic wave-packet, every point x in space has its own clock, giving the τ for that point, leading to complicated relations between k and ω.

From Eq. 6.1, we can determine $a(k)$ as

$$a(k) = \frac{1}{\sqrt{2\pi}} \int_{-\infty}^{\infty} u(x, 0)e^{-ikx}dx \tag{6.2}$$

For a Gaussian wave-packet at $\tau = 0$,

$$u(x, 0) = e^{-\sigma^2 x^2 + ik_0 x} \tag{6.3}$$

Substituting from Eq. 6.3 into Eq. 6.2 we get

$$a(k) = \frac{1}{\sigma\sqrt{2}} e^{-\frac{(k-k_0)^2}{4\sigma^2}} \tag{6.4}$$

Therefore, from Eqs. 6.1 and 6.4 we get

$$u(x, \tau) = \frac{1}{\sqrt{2\pi}} \int_{-\infty}^{\infty} dk \frac{1}{\sigma\sqrt{2}} e^{\frac{-(k-k_0)^2}{4\sigma^2}} e^{i(kx-\omega\tau)} \tag{6.5}$$

At the junction (x_i, τ_i) of the source reservoir and lead, we can write

$$u(x_i, \tau_i) = \frac{1}{\sqrt{2\pi}} \int_{-\infty}^{\infty} dk \frac{1}{\sigma\sqrt{2}} e^{\frac{-(k-k_0)^2}{4\sigma^2}} e^{i(kx_i-\omega\tau_i)} \tag{6.6}$$

Since the source in Landauer-Buttiker formalism is classical, (x_i, τ_i) is consistent with the laws of relativity. At the junction of the output lead with the sink reservoir

$$u_{tr}(x_i + L, \tau_i + \tau_0 + \Delta\tau_0) =$$

$$\frac{1}{\sqrt{2\pi}} \int_{-\infty}^{\infty} dk \frac{1}{\sigma\sqrt{2}} t(k) e^{\frac{-(k-k_0)^2}{4\sigma^2}} e^{ik(x_i+L)-i\omega(\tau_i+\tau_0+\Delta\tau_0)} \tag{6.7}$$

Here L is the distance between the two reservoirs, $t(k)$ is transmission amplitude of the length L, and τ_0 is the time taken to transmit the wave-packet in the absence of potential in the region of length L, i.e., when $t(k) = 1$. We know that in absence of any potential as well as in the semi-classical limit, the wave-packet remains unchanged in shape but broadens. Note from Eq. 1.96 that there are two conventions and in the other convention one has to include the factor e^{ikL} in $t(k)$ and thereby also remove τ_0 in the exponential term in Eq. 6.7. In the convention of Eq. 6.7, $\tau_0 + \Delta\tau_0$ is the time taken to transmit the wave-packet (assume undispersed or limited dispersion) in the presence of potential. Hence

$$u_{tr}(x_i + L, \tau_i + \tau_0 + \Delta\tau_0) =$$

$$\frac{1}{\sqrt{2\pi}} \int_{-\infty}^{\infty} dk \frac{1}{\sigma\sqrt{2}} |t(k)| e^{i\eta_k} e^{\frac{-(k-k_0)^2}{4\sigma^2}} e^{ik(x_i+L)-i\omega(\tau_i+\tau_0+\Delta\tau_0)} \tag{6.8}$$

where $t(k) = |t(k)|e^{i\eta_k}$. Now for the weight factor $\frac{1}{\sigma\sqrt{2}}|t(k)|e^{\frac{-(k-k_0)^2}{4\sigma^2}}$, Eq. 6.8 still represents a localized wave-packet (localized in k space implies localized in real space, a property that is well captured by Fourier transforms) at the junction of the sink reservoir at time $\tau_i + \tau_0 + \Delta\tau_0$, its amplitude peaking in k-space at k_0 and exponentially decaying on either sides. Since the sink reservoir is classical, $\tau_i + \tau_0 + \Delta\tau_0$ is the time when the wave-packet reaches the sink reservoir. For a classical pulse, the length L and the speed cum acceleration of the pulse will determine the propagation time which we may call $\tau_0^c + \Delta\tau_0^c$, where the superscript "c" stands for "classical", and this problem is completely in tandem with the theory of relativity that gives a proper time and a coordinate time. Now for a quantum mechanical electron pulse in Eq. 6.8, $\tau_0 + \Delta\tau_0$ is just a shift of origin of the time axis, implying the wave-packet has shifted along the time axis, while the shift in spatial axis is $L + \frac{\eta_k}{k}$. The only two system dependent terms being $\Delta\tau_0$ and η_k, they come with signs that can make $\Delta\tau_0$ negative that never happen for classical systems. In analogy with optics, $\frac{\eta_k}{k}$ can be referred to as an optical path that is the physical path traversed effectively slow or fast (a mechanism different from acceleration). Slow is intuitively clear but there are questions about the fast that we address in this chapter. For classical electromagnetic waves multiple reflection is a physical phenomenon but that is not the case for quantum mechanics. In quantum mechanics we talk of multiple reflection in terms of Feynman paths and not physical paths. In the canonical approach wave-functions are not waves but rather functions and functions do not get reflected. Besides the

LPT approach reveals that there is a unique local time at every point in space and not that the particle runs through the same point repeatedly as it happens for classical waves undergoing multiple reflections. It is known in the von-Neumann approach that geometrically one needs to classify wave-functions with respect to bundles. If σ^2 is very small then in k space the wave-packet is sharply peaked around k_0 and in this range $|t(k)|$ can be taken to be independent of k to help pinpoint the real issues and so $|t(k)| = t$. The real issue is that even for a k dependent transmission, Eq. 6.8 can be a localized wave-packet of a predictable shape but the k dependent transmission phase shift can disperse it because that affects an infinite number of components in the integration. A detector at $x = L$ usually detects the following Fourier transform (dropping the constant factors):

$$\int_{-\Delta\tau_d}^{\Delta\tau_d} d\tau e^{-i\omega\tau}\left[\int_{-\infty}^{\infty} dk|t(k)|e^{i\eta_k} e^{\frac{-(k-k_0)^2}{4\sigma^2}} e^{ik(x_i+L)-i\omega(\tau_i+\tau_0+\Delta\tau_0)}\right] \tag{6.9}$$

where $2\Delta\tau_d$ is the detection time interval, which is like the shutter speed for a perfectly 2D camera sensor located at $x = L$ and operated at $\tau_i + \tau_0 + \Delta\tau_0$. So the detector in fact detects the components in k space (or ω space) between times $\tau_i + \tau_0 + \Delta\tau_0 - \Delta\tau_d$ and $\tau_i + \tau_0 + \Delta\tau_0 + \Delta\tau_d$ that peak at k_0 but the the shift along spatial axis is $L + \frac{\eta_k}{k}$. Thus the detector actually detects a peak in k space or in ω space and so a sharply peaked wave-packet in k space is not a problem. If dispersion is not much then in real space it is a localized pattern, propagating in time and so can carry a signal. For too strong dispersion or imperfect detectors, the peak will be washed out but within time scale experiments with electro-magnetic wave-packets as well as quantum wave-packets show these peaks and form the basis for a major subject called spectroscopy. So $\tau_0 + \Delta\tau_0$ will be clocked by the local classical clock at the junction of sink reservoir and lead which is an interval of time since τ_i, and this interval will be independent of τ_i and also different from $\tau_0^c + \Delta\tau_0^c$ (which is always positive). It may seem that we demand too much fine tuning from an experimentalist. So we will argue that if a slightly broadening quantum wave-packet can be used to carry a signal in positive time, which is an established fact in 1D, then at a Fano resonance a wave-packet can carry a signal in negative time. We will talk about some much simpler experiments in the next chapter, one of which bypasses the quantum measurement problem as well as the way in which detection is perceived in spectroscopy, and the other is a well established experimental method in mesoscopic physics. From Eq. 6.8, with $|t(k)| = t$

$$u_{tr}(x_i + L, \tau_i + \tau_0 + \Delta\tau_0) =$$

$$\frac{1}{2\sqrt{\pi}}\frac{1}{\sigma}t\int_{-\infty}^{\infty} dk e^{\frac{-(k-k_0)^2}{4\sigma^2}} e^{i\eta_k+ik(x_i+L)-i\omega(\tau_i+\tau_0+\Delta\tau_0)} \tag{6.10}$$

So for the wave-packet to remain undispersed the weight of a particular component must remain unchanged implying the arguments of the exponential terms in Eqs. 6.6 and 6.10 should lead to

$$kx_i - \omega\tau_i = \eta_k + k(x_i + L) - \omega(\tau_i + \tau_0 + \Delta\tau_0) = \text{constant} \tag{6.11}$$

$$\text{or,}\ \ \eta_k + kL - \omega(\tau_0 + \Delta\tau_0) = 0$$

$$\text{or,}\ \ \frac{d\eta_k}{d\omega} + L\frac{dk}{d\omega} = \tau_0 + \Delta\tau_0 \tag{6.12}$$

Also in the absence of scatterer

$$\therefore\ \ kx_i - \omega\tau_i = k(x_i + L) - \omega(\tau_i + \tau_0)$$

$$\text{or,}\ \ kL - \omega\tau_0 = 0$$

$$\text{or,}\ \ L\frac{dk}{d\omega} = \tau_0 \tag{6.13}$$

Therefore from Eq. 6.12

$$\frac{d\eta_k}{d\omega} = \Delta\tau_0$$

$$\frac{d\eta_k}{dE}\frac{dE}{d\omega} = \Delta\tau_0$$

$$\hbar\frac{d\eta_k}{dE} = \Delta\tau^W \qquad \text{as}\ \ E = \hbar\omega \tag{6.14}$$

That is the $\Delta\tau_0$ defined in Eq. 6.7 and calculated by imposing stationary phase approximation on the propagating wave-packet represented by Eq. 6.7 is called $\Delta\tau^W$ or Wigner delay time (WDT) given by Eq. 6.14. Since any scattering phenomenon can be quantified through partial wave analysis or equivalently by time evolution of a wave-packet, when WDT is correct, it has to give the same physical quantity as LPT. As shown in next section it is expected at the phase drops and so in Figs. 4.4 and 5.11 the semi-classical formula and hence WDT becomes exact at the Fano resonances. Figure 5.13a specially shows the fact that the exactness happens at energy of the phase drop that is slightly greater than the peak in DOS, which can be also seen in Figs. 4.4 and 5.11 if resolved properly. Thus if a signal can be sent in WDT (the correct one) it can be also sent in LPT. In the following we will quantify this analysis further and show that it will be true for any potential that give a Fano resonance because it depends on the only fact that we get a completely closed sub-loop within one Riemann surface. We will also discuss how the stationary phase approximation plays its role.

6.4 Novelty of Q1D

Let us now go beyond 1D where the S-matrix is 2×2 and consider the three prong potential of Fig. 5.6 for which the S-matrix is 3×3. We get $k = \sqrt{\frac{2m_0 E}{\hbar^2}}$ and $q = \sqrt{\frac{2m_0(E-U)}{\hbar^2}}$ are wave vectors along the thin lines and the thick solid lines, respectively. An incident wave-packet along lead 1 will split into three wave-packets that are reflected and transmitted to leads 2 and 3. Once again, according to quantum mechanics, if we project one particle or a small number of particles into a scatterer, represented by a wave-packet propagating in time, then the particles may get reflected or transmitted to 2 or to 3 in a random manner. But if we repeat the process with a statistically large number of electrons then it will produce a reproducible interference pattern and all observable quantities like r_{11}, t_{21}, and t_{31} that can be obtained from the stationary state solutions of the time-independent Schrodinger equation. In other words, η_k in Eq. 6.14 or $\theta_{t_{31}}$, $\theta_{t_{21}}$ of Fig. 5.7 is in practice always obtained from the time-independent Sc. equation. Identical analysis allows us to treat $\theta_{t_{21}}$ of Fig. 5.7b in the same footing as η_k. The WDT $\Delta\tau_{21}^W$ for the wave-packet coming out of lead 2 will be

$$\Delta\tau_{21}^W = \hbar \frac{d\theta_{t_{21}}}{dE} \tag{6.15}$$

In Fig. 5.7b we can see the typical semi-classical behavior like in 1D wherein $\theta_{t_{21}}$ is monotonously rising with a point of inflection signifying a peak in $\hbar\frac{d\theta_{t_{21}}}{dE}$ that corresponds to a peak in PDOS for those partial electrons going from lead 1 to lead 2. This semi-classical behavior is well known and can be quantified by the fact that the de-Broglie wavelength is smaller than the spatial scale in which the potential varies. It is generally known that for this kind of phase behavior, the stationary phase approximation is valid within a time scale, WDT gives the correct signal propagation time which is always positive and in agreement with LPT summed over coordinates. LPT summed over coordinates essentially gives a traversal time of electron propagation from reservoir 1 to reservoir 2 as

$$\Delta\tau_{lpt}(E, 2, 1) = -\hbar \int_{sample} d^3\mathbf{r} \frac{\delta\theta_{t_{21}}}{\delta U(\mathbf{r})} \tag{6.16}$$

For this case, it is no surprise that $\Delta\tau_{21}^W = \Delta\tau_{lpt}(E, 2, 1) > 0$ and it can be easily checked for the three prong potential. Similarly, the WDT $\Delta\tau_{31}^W$ for the wave-packet coming out of lead 3 will be

$$\Delta\tau_{31}^W = \hbar \frac{d\theta_{t_{31}}}{dE} \tag{6.17}$$

and this has complications. As discussed in Chap. 5, $\theta_{t_{31}}$ in Fig. 5.7d does not show the typical behavior as seen for $\theta_{t_{21}}$ in Fig. 5.7b, because of the presence of phase drops where $\Delta\tau_{31}^W = \hbar\frac{d\theta_{t_{31}}}{dE}$ becomes negative. Therefore for this case one can question if LPT integrated over coordinates becomes identically negative. If yes then the validity of stationary phase approximation should be straightforward conclusion from Copenhagen school, but we will also analyze that in more detail. It is discussed in Chap. 5 (see the paragraph after Eq. 5.23) that the negative slope in Fig. 5.7d at the point marked P is also seen in 1D scattering, is well studied, and violates stationary phase approximation. So one can state that for the three prong scattering, a wave-packet coming out of lead 3 in this regime, WDT and LPT averaged over coordinates disagree (can be seen in Fig. 6.5), and WDT is not consistent with DOS and is hence wrong. According to our analysis, this is due to the fact that this phase drop does not come from closed sub-loops of a Fano resonance and therefore it can be also seen in 1D. The phase drop of the type at the point Q is new and a consequence of Fano resonance that results in a closed sub-loop within one Riemann surface. We will show that for this case WDT and LPT both lead to identical negative times. Although we will calculate this for $\theta_{t_{31}}$ of the three prong potential, we will show that it is a consequence of the topology of the complex plane obtained from imaginary part of the scattering matrix element versus the real part and is a general feature of Fano resonances coming from any potential scattering. Before we discuss the drops in $\theta_{t_{31}}$, it is suitable to consider θ_t of Fig. 5.3c that show how a discontinuous phase drop evolves into a sharp but continuous phase drop as that exhibited by $\theta_{t_{31}}$.

Our interest is in the dashed curve in the range of k values around the point M in Fig. 5.3c, in which θ_t show a linear drop. It originates from the transmission zero of the Fano resonance that is separate from the transmission pole position, has no point of inflection, and is hence linear. If we construct a wave-packet with k values in a narrow range around point M, then for Eqs. 6.10 and 6.11, $\eta_k \equiv \theta_t$ decrease linearly around zero and the stationary phase approximation can be imposed. To see that consider Eq. 6.10 and dropping the constant factors we get

$$u_{tr} = \int_{-\infty}^{\infty} dk e^{\frac{-(k-k_0)^2}{4\sigma^2}} e^{ik\frac{\eta_k}{k}+ik\left(x_i+L\right)-i\omega\left(\tau_i+\tau_0+\Delta\tau_0\right)} \tag{6.18}$$

$$u_{tr} = \int_{-\infty}^{\infty} dk e^{\frac{-(k-k_0)^2}{4\sigma^2}} e^{+ik\left(x_i+L+\frac{\eta_k}{k}\right)-i\omega\left(\tau_i+\tau_0+\Delta\tau_0\right)} \tag{6.19}$$

To test the validity of stationary phase approximation, let us see if that condition can be satisfied for k_0 at the point M in Fig. 5.3c of the dashed curve.

$$u_{tr} = \int_{-\infty}^{\infty} dk e^{\frac{-(k-k_0)^2}{4\sigma^2}} e^{+ik\left(x_i+L+\frac{\theta_t}{k}\right)-i\omega\left(\tau_i+\tau_0+\Delta\tau^W\right)} \tag{6.20}$$

From Eq. 6.14

$$u_{tr} = \int_{-\infty}^{\infty} dk e^{\frac{-(k-k_0)^2}{4\sigma^2}} e^{+ik\left(x_i+L+\frac{\theta_t}{k}\right)-i\omega\left(\tau_i+\tau_0+\hbar\frac{d\theta_t}{dE}\right)} \tag{6.21}$$

Since around the point M in Fig. 5.3c of the dashed curve, θ_t decreases linearly with k, therefore $\frac{\theta_t}{k}$ as well as $\frac{d\theta_t}{dE}$ are constants and such constant phase shifts do not affect the shape of u_{tr} for small σ. Moreover at the point M, $\theta_t = 0$. Note the fact that θ_t and $\frac{d\theta_t}{dE}$ are the only system dependent terms.

Traversal time that one gets from LPT for the electrons going from lead 1 to lead 3 of the three prong scattering is

$$\Delta\tau_{lpt}(3, 1, E) = -\hbar \int_{sample} d^3\mathbf{r} \frac{\delta\theta_{t_{31}}}{\delta U(\mathbf{r})} \tag{6.22}$$

Note that $|t_{31}|^2$ fraction of the incident electrons gets transmitted like this. As usual, $\frac{\delta}{\delta U(\mathbf{r})}$ means a functional derivative with respect to the local potential $U(\mathbf{r})$ and $\int_{sample} d^3\mathbf{r}$ means an integration over all the coordinates of the sample or the scattering region. Definition of the scattering region is the spatial region where the wave-function deviates from the typical asymptotic wave-functions of a scattering problem. In case of Fig. 5.6 the sample or the scattering region is simply the region shown in thick lines. This means that to calculate the functional derivative for the three prong potential, we change $U(\mathbf{r})$ at all points inside the sample by an infinitesimal amount of $e_0\Delta\epsilon$ where e_0 is the incident particle charge. Since the differential increment is independent of $\mathbf{r}$, in depicting the figures we will not differentiate between derivative and functional derivative. For realistic $\mathbf{r}$ dependent potentials or even for the two channel delta function potential of Chap. 4, this functional derivative is almost impossible to calculate, but the properties of the AD make it evident that our conclusions will be valid for them.

To summarize, we would first like to clearly define the three terms we use subsequently, in this chapter. First is the "traversal time" which will be used in a literary sense that it is a time associated with a signal going from source reservoir to sink reservoir traversing the intermediate states connecting the two points. So for a time interval to qualify as traversal time it has to be related to DOS that constitutes the signal. Second is the "LPT" which is exact and always sums up to give the correct DOS. Third is the "WDT" which is correct if stationary phase approximation is correct and not otherwise. However, from now on we will not use terms like "correct WDT" and "incorrect WDT". Correctness or incorrectness will be born by its comparison with LPT. If WDT is correct then it has to be equal to LPT. Next we will analyze the situation for the phase drop at the point marked P in Fig. 5.7d that reveals the fact that in 1D at low energy, LPT remains positive while WDT may become negative. We use Burgers circuit (BC) to prove this. Next we will show that in Q1D, one can find frequently occurring phase drops like the one at the point marked Q where LPT and

WDT can both be identically negative. The essential difference for the phase drop at Q with the one at P is that the drop at Q originates from a closed sub-loop within one Riemann surface. That is we will prove that in Q1D there are robust regimes where we get phase drops of the type at point Q for which

$$\Delta\tau_{lpt}(3, 1, E) = \Delta\tau_{31}^{W} < 0 \tag{6.23}$$

Before showing this we would like to state why such a demonstration is interesting. First of all it is new and no such previous examples or a system that exhibit this can be cited. Secondly it implies that $\Delta\tau_{lpt}(3, 1, E)$ can be negative and such a quantitative negative value for it has not been obtained before except in simple ideal 1D situations. $\Delta\tau_{lpt}(3, 1, E) < 0$ does not prove anything with respect to signal propagation time unless one can show $\Delta\tau_{lpt}(3, 1, E) = \Delta\tau_{31}{}^{W} < 0$ because only then one can claim that a wave-packet or some change in spatial probability distribution of electrons can be transmitted to negative times within the single particle coherence length. Since quantum mechanics and relativity are often not consistent with each other but both very successful in their own regimes it has been always believed that superluminal times are possible in quantum mechanics. But a concrete demonstration consistent with quantum mechanics itself has not been shown so far. We will use the three prong potential to illustrate our results. The proof is general and can be applied to any scattering matrix element $s_{\alpha\beta}$ whose AD make loops within one Riemann surface and in mesoscopic systems one can find a large class of systems that exhibit this essentially due to Fano resonance. Burgers circuit applied to AD uses the analyticity of the complex scattering matrix elements and is therefore more general than quantum mechanics and so quantum mechanics has to respect it.

In Fig. 6.1 we give a part of the Argand diagram in Fig. 5.7c which is the AD for t_{31} and in Fig. 6.2 similarly we give a part of Fig. 5.7d which is a plot of $\theta_{t_{31}}$ as a function of the wave vector for the same energy range as in Fig. 6.1. Note that in Fig. 6.2 around the point marked P, the scattering phase shift decreases with kl

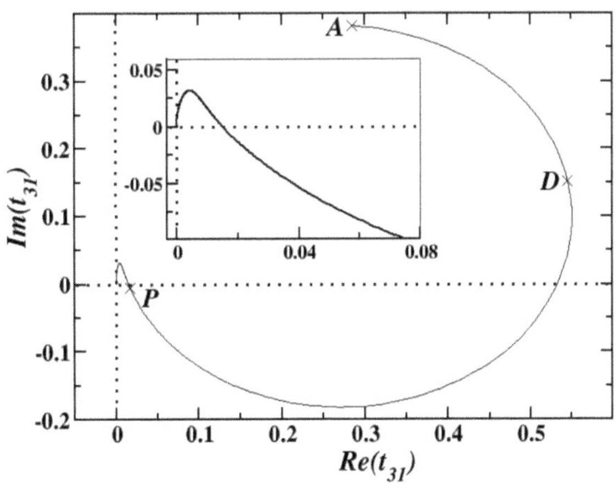

Fig. 6.1 Argand diagram for transmission amplitude t_{31} for the three prong potential shown in Fig. 5.6. The Argand diagram is obtained by varying the wave vector kl from 0 to 5. Here, $l_1 = l_3 = l, l_2 = 5l$, and $e_0Ul = -1000$. The inset shows a magnified picture of the region near the origin

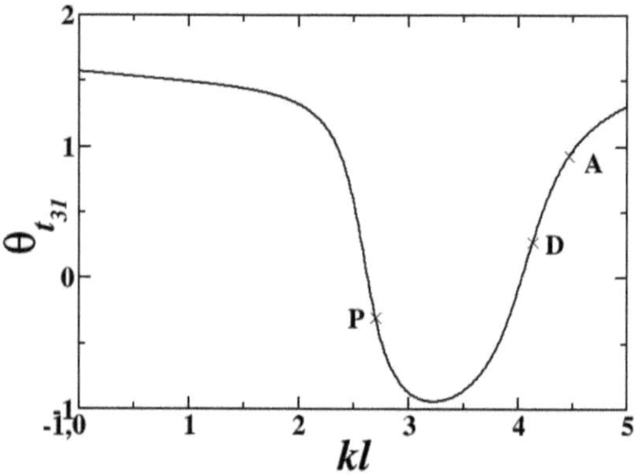

Fig. 6.2 Plot of transmission phase shift $\theta_{t_{31}}$ corresponding to transmission amplitude t_{31} for the three prong potential shown in Fig. 5.6, as a function of kl. Here, $l_1 = l_3 = l, l_2 = 5l$, and $e_0Ul = -1000$

which means $\frac{d\theta_{t_{31}}}{dE}$ $(= \frac{m_0 l}{\hbar^2 k}\frac{d\theta_{t_{31}}}{d(kl)})$ will be negative. The corresponding point in the AD is also marked P in Fig. 6.1. One can calculate $\theta_{t_{31}} = Arctan[\frac{Im(t_{31})}{Re(t_{31})}]$ at point P in Fig. 6.1 at close by points to see from the Argand diagram as well that $\theta_{t_{31}}$ has a negative slope here. This negative slope does not have anything to do with the three prong potential and it can be observed for all values of l_2 at the same energy and of similar magnitude. For example by making $l_2 \to 0$ in Fig. 5.6, one will get the 1D limit where too one can check and find this negative slope at the same energy and of similar magnitude. This negative slope in 1D has been the matter of study in the past with respect to traversal time, and there is no conclusive understanding as to whether it means something physical or is an artifact of the stationary phase approximation [2]. It can be seen in a one-dimensional potential like a square barrier or a square well potential ($l_2 \to 0$ limit of Fig. 5.6) where it happens only once as the energy or the momentum is increased from 0. We can understand the origin of this negative slope from the AD using BC. At zero energy the transmission is zero even for an infinitesimal potential in the scattering region, implying $Im(t_{31})$ and $Re(t_{31})$ are both zero. This is a point where $\theta_{t_{31}}$ is singular. So the AD starts or originates from the singularity and comes out of the singularity as the incident energy increases from 0. Now since the absolute value of t_{31} is bounded the AD has to curl around the singularity either in clockwise direction or anticlockwise direction. Clockwise will mean phase will decrease whereas anticlockwise will mean phase will increase with energy as suggested by Eq. 5.1. Initially, however, when the AD trajectory emerges from the singularity, it is neither clockwise nor anticlockwise with respect to it. After a while it turns around and before becoming anticlockwise with respect to the origin it goes through a small region around point P where it is clockwise, and this is what creates the negative slope. Once it has gone anticlockwise it never turns back again in 1D. And so after this initial drop phase monotonously increases. The phase of the wave-function increases implies rotation of the wave-function in Hilbert space unidirectionally which is like a notion of time increasing monotonously and positively, as time evolution is guided by a factor $e^{iEt/\hbar}$. Now in case of the three prong potential we would like to make the following observations for the phase drop

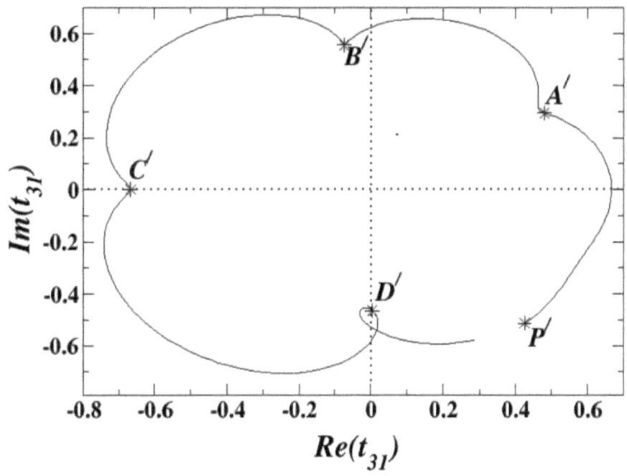

Fig. 6.3 Argand diagram for transmission amplitude t_{31} for the three prong potential shown in Fig. 5.6. The Argand diagram is obtained by varying the potential e_0Ul from 0.0 to -25. It starts from the point marked $P^/$. Here, $l_1 = l_3 = l$, $l_2 = 5l$, and $kl = 2.7$ which are the parameters corresponding to the point P in Figs. 6.2 and 6.3. Even when kl value is taken close to zero rather than 2.7, the point $P^/$ does not get close to the origin. The phase singularity is at the origin where $t_{31} = 0$

at P, that for some reason has not been observed in 1D essentially because making these observations is not very revealing in 1D. Our observations for the three prong potential will become more relevant as the chapter progresses.

First of all one can keep the energy (or kl) fixed at the value corresponding to the point P in Fig. 6.1 and decrease (or increase) the potential globally by constant amounts to generate identically same AD starting from P. Energy increasing corresponds to potential decreasing globally and vice verse. But this is a very obvious observation. Now to come to the point, suppose we keep the energy fixed at that corresponding to point P and decrease the potential only in the thick region in Fig. 5.6 (that is only in the scatterer) by constant amounts starting from 0. Unlike the AD of Fig. 6.1 this AD shown in Fig. 6.3 will not start from the singularity but start from a point completely different from P. Because when the potential is zero, even a mode with infinitesimal energy or momentum gets fully transmitted. So even if the point P in Fig. 6.1 is taken arbitrarily close to the origin, the AD obtained by varying the potential U (in the thick region of Fig. 5.6, starting from $U = 0$) will not start from the origin. It starts from P$^/$ and right from the very start it goes anticlockwise with respect to the origin or with respect to the singularity, its trajectory being completely unaffected by the presence of the singularity. Only when a trajectory emerges from the origin, it can neither be clockwise or anticlockwise with respect to the origin. The scattering phase shift shown in Fig. 6.4 corresponding to this AD shown in Fig. 6.3 therefore starts increasing from –1 and there is no negative slope near the origin. Only at a higher value of the potential U, i.e., at the point D$^/$ it shows a negative slope due to the small sub-loop at $D^/$ in Fig. 6.3. Decreasing the constant potential U only in the thick region of Fig. 5.6 in small steps of $e_0\Delta\epsilon$ allows us to

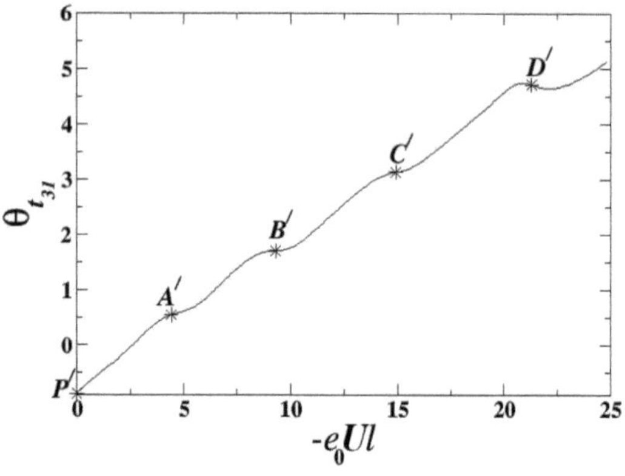

Fig. 6.4 Plot of transmission phase shift $\theta_{t_{31}}$ corresponding to the AD in Fig. 6.3, as a function of $-e_0Ul$ (U is negative and decreasing)

evaluate the RHS of Eq. 6.22 and thus $\Delta\tau_{lpt}(3, 1, E)$. Here e_0 is the particle charge that is set to 1 without any loss of generality. Thus AD explains why LPT will not be negative at P and can be verified from Eq. 6.22. So at low energies or in the quantum regime, global constant shift in potential and constant shift of potential in the scattering region produce completely different results. Adding up all possible LPT for all the channels like t_{21}, r_{11}, etc., and multiplying with $\frac{1}{h}$ one can get density of states (DOS) that can be independently determined from the internal wave-function also for comparison. We have checked that there is perfect agreement of this LPT calculated from Eq. 6.22 with the DOS determined from internal wave-function. In Fig. 6.5 we plot $\frac{d\theta_{t_{31}}(E)}{dE}$ and $\frac{d\theta_{t_{31}}(E)}{d(e_0U)}$ where U is the constant potential in the sample region, i.e., the region shown in Fig. 5.6 in bold lines. As mentioned before, for a constant U, we are denoting functional derivatives as derivatives. Both are plotted as a function of kl. It shows $\frac{d\theta_{t_{31}}}{dE}$ is strongly negative near $k = 0$ but $\frac{d\theta_{t_{31}}}{d(e_0U)}$ is not. This negativity of $\frac{d\theta_{t_{31}}}{dE}$ is due to the failure of stationary phase approximation as it does not agree with LPT. LPT starts from zero and starts going negative from the very start (note that both the derivatives in Fig. 6.5 are calculated at a large fixed value for potential U in the sample). However, since LPT does not agree with WDT Eq. 6.23 is not satisfied. However, at three broad minima shown in Fig. 6.5 and mentioned in the figure caption, Eq. 6.23 is satisfied. There are other regions also where $\frac{d\theta_{t_{31}}}{dE}$ and $\frac{d\theta_{t_{31}}}{d(e_0U)}$ are identical but since they are not negative there Eq. 6.23 is not satisfied.

Now in case of Fig. 6.1, if we increase the energy further then the AD show sub-loops within a particular Riemann surface and demonstrated in Fig. 6.6. At each sub-loop we get a portion of AD trajectory that is clockwise with respect to the origin. For example we have highlighted one such sub-loop in Fig. 6.6 by a dotted circle and marked the sub-loop as MNOM. Between N and O the trajectory is clockwise or in other words the normal vector to the curve points towards the origin and scans the origin. Between N and O, $\frac{d\theta_{t_{31}}}{dE}$ will be negative. The identical AD of Fig. 6.6 can be generated by keeping the energy fixed at the value corresponding to point M (say) and decreasing (or increasing) the potential globally by constant amounts. We

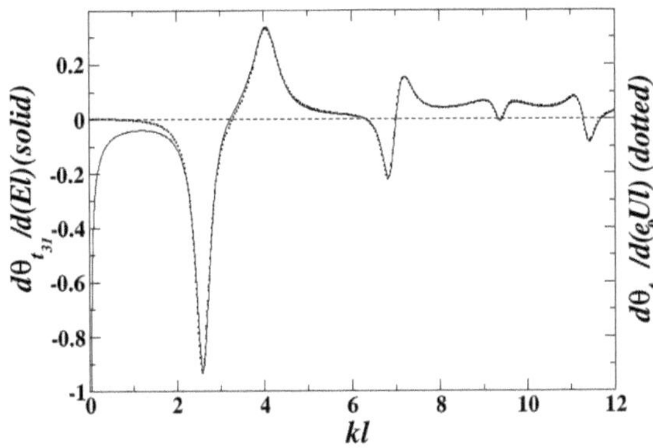

Fig. 6.5 Here, $l_1 = l_3 = l, l_2 = 5l$, and $e_0Ul = -1000$. Around three broad minima at $kl = 2.5, 6.7$, and 11.4, Eq. 6.15 is satisfied. For a constant scattering potential in the sample, we denote functional derivative as derivative

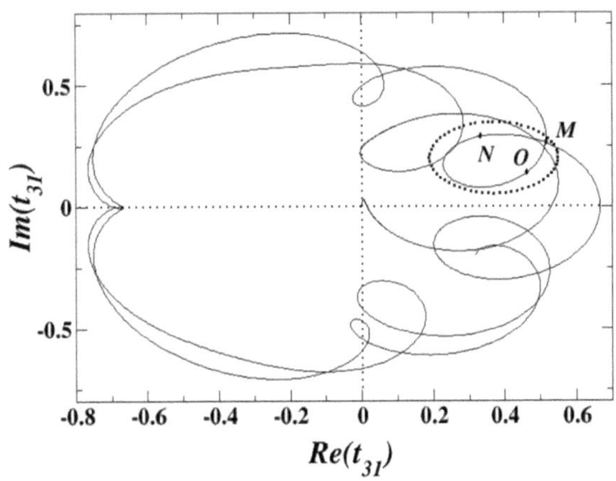

Fig. 6.6 Argand diagram for transmission amplitude t_{31} for the three prong potential shown in Fig. 5.6. kl is varied from 0 to 20, to get this AD. Here, $l_1 = l_3 = l, l_2 = 5l$, and $e_0Ul = -1000$. The dotted circle is just to highlight the sub-loop $MNOM$ in the AD

have already discussed that in a scattering problem increasing incident energy by dE is equivalent to decreasing the potential globally by a constant amount $\Delta\varepsilon$, such that $dE = -e_0\Delta\varepsilon$, where e_0 is particle charge that we will set to 1 to simplify our arguments. That is, the changed potential is $U'(\mathbf{r}) = U(\mathbf{r}) - \Delta\varepsilon$ for $-\infty < \mathbf{r} < \infty$. Hence if we can generate a closed sub-loop in the Argand diagram by varying E in small steps of dE, then we can also do so by globally changing the potential in small steps of $\Delta\epsilon$. Decreasing the potential globally is equivalent to increasing the energy and so it will generate the AD further away from the origin (starting from M and towards N) and increasing the potential is equivalent to decreasing the energy and so it will generate the AD from M down to the origin. So the sub-loops can also be generated by decreasing the potential globally (that includes the thick regions as well as in the leads in Fig. 5.6) starting from the U value used in Fig. 6.6 and keeping the energy fixed at the value corresponding to point M in Fig. 6.6. For a closed contour A corresponding to a particular sub-loop (say MNOM) in Fig. 6.6,

$$\oint_A \Delta\theta_{s_{\alpha\beta}} = 0$$

$$\text{or } \oint_A \frac{\Delta\theta_{s_{\alpha\beta}}}{\Delta E} dE = 0 \tag{6.24}$$

where increment in E, that is $\Delta E \to 0$. This sub-loop A can also be generated by a global appropriately large shift of potential in infinitesimal steps. These sub-loops are consistent with BC and hence the topological structure of the Riemann surface in which the AD is drawn. Sub-loops also belong to a certain topological class such that a particular sub-loop is very fundamental and cannot be removed by varying any parameter (note that in previous chapter we said that presence or absence of a sub-loop does not have any effect on the conserved quantity I in Eq. 5.1 and so sub-loops can occur without violating any principle. Now we are using topological argument to say that once we identify a sub-loop, it cannot be removed by adjusting any parameter, where of course we are excluding possibilities of changing the system itself like jumping from a closed to an open system. A sub-loop can grow or reduce in size continuously and can even become infinitesimal when it appears like a cusp). Essentially it means a multiply connected curve cannot be continuously deformed to a simply connected curve by tuning terms and parameters in the Hamiltonian (there are no phase transitions in small quantum systems and in the regime of single particle coherence length we are anyway not concerned with them). Physically, these sub-loops originate from Fano resonance and a particular resonant state cannot be removed by varying any parameter. A resonant state can be populated or de-populated but the total number of states in a system remains conserved. So if we do not increase the incident energy (or decrease the potential globally) but decrease the potential only in the thick region of Fig. 5.6 then we will generate identical number of sub-loops. The nature of the different sub-loops may change but the total number of sub-loops will be the same. This is shown in Fig. 6.7 up to very high values of the potential in the sample region (such high energies are not shown in Fig. 6.6). Therefore, for a closed contour A in Fig. 6.6, one can find the corresponding closed contour B in Fig. 6.7 (essentially one has to find the sub-loop B coming from a Fano resonance with the same number of nodes as the Fano resonance making sub-loop A), such that

$$\oint_B \delta\theta_{s_{\alpha\beta}} = -\oint_B \Delta\varepsilon \int_{sample} \frac{\delta\theta_{s_{\alpha\beta}}}{\delta U(\mathbf{r})} d^3\mathbf{r} = 0 \tag{6.25}$$

By the argument that if a sub-loop A can be generated by varying the incident energy E then a corresponding sub-loop B can be generated by varying the potential uniformly in the sample region only, in small steps of $e_0\Delta\epsilon$ and all sub-loops can be thus paired up giving (from Eqs. 6.24 and 6.25)

$$\oint_A \frac{\Delta\theta_{s_{\alpha\beta}}}{\Delta E} dE = -\oint_B \Delta\varepsilon \int_{sample} \frac{\delta\theta_{s_{\alpha\beta}}}{\delta U(\mathbf{r})} d^3\mathbf{r} = 0 \tag{6.26}$$

This only proves that the integrations are equal. This also proves that the integrands are negative in part of the range in which the relevant parameter is varied to get the closed contours A and B, and positive in the rest of it. Thus this proves that LPT

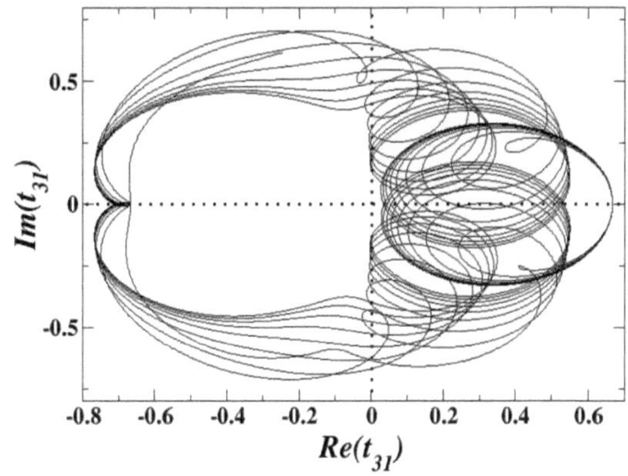

Fig. 6.7 Argand diagram for transmission amplitude t_{31} for the three prong potential shown in Fig. 5.6. The Argand diagram is obtained by varying the potential e_0Ul, that is the potential in the sample region, from -1 to -1000. Here, $l_1 = l_3 = l$, $l_2 = 5l$, and $kl = 4$

which is always correct can be negative. All the sub-loops A in Fig. 6.6 and pairing sub-loops B in Fig. 6.7 will give energy ranges in which Eq. 6.26 is satisfied. But this does not explain that around the three minima in Fig. 6.5 at negative values why Eq. 6.23 should be valid. In Fig. 6.5 we are plotting the integrands in Eq. 6.26. To explain that we have to show that the integrands in Eq. 6.26 can be equal in some regimes. If we can find some regime where the value of the integration becomes independent of the shape and size of the contour A as well as B, then the integrands have to be identical. And also unlike the sub-loops in Figs. 6.6 and 6.7 the trajectory making sub-loops A and B should not have any discontinuities but should come back onto itself very smoothly or else the integration limits (like the point M in Fig. 6.6 in the trajectory M to N is not identical to the point M in the trajectory O to M) are not identical in all respect. If closed contour A satisfies these two properties then the pairing closed contour B too has to satisfy them. Because the LHS integration cannot be 0 independent of the shape and size of the contour A while the RHS integration does depend on the shape and size of the contour B. If the end points are identical, then two analytic functions are identical in between. Such smooth sub-loops A are shown in Fig. 6.8 and smooth sub-loops B are shown in Fig. 6.9. Relevant parameters are mentioned in the figure caption. Note that in Fig. 6.9 we have fixed the energy such that $kl = 8.22$ and the initial value of $e_0Ul = -1000$ which implies the trajectory will start exactly from the point marked x in Fig. 6.9, which is identical to the point at $kl = 8.22$ in Fig. 6.8. The purpose of Fig. 6.8 is to show that by making $l_1 = l_3 = 0$ in Fig. 6.6 we can make the sub-loops in Fig. 6.6 become smoothly coming back to one particular point. Similarly the purpose of Fig. 6.9 is to show that by making $l_1 = l_3 = 0$ in Fig. 6.7 we can make the sub-loops in Fig. 6.7 become smoothly coming back to one particular point.

In Fig. 6.10 we plot $\frac{d\theta_{t_{31}}(E)}{dE}$ and $\frac{d\theta_{t_{31}}(E)}{d(e_0U)}$ as a function of kl and above a value of $kl = 8.22$ the two curves become identical as expected from Figs. 6.8 and 6.9. There are broad energy ranges occurring periodically where both curves become identically negative and so Eq. 6.23 is satisfied and this happens up to very high energies. If we construct a wave-packet with the modes in these ranges then a wave-packet will

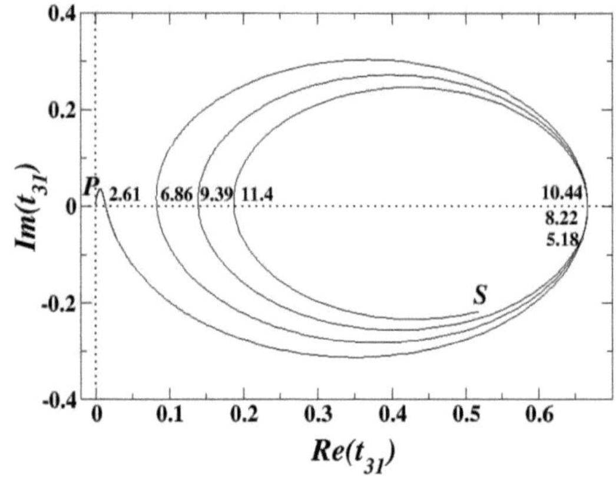

Fig. 6.8 AD for t_{31} for the system as in Fig. 6.6 for the same parameter values as that in Fig. 6.6, but with $l_1 = l_3 = 0$. kl is varied from 0 to 12.0 and some kl values like 2.61, 5.18, 6.86, etc., are marked at corresponding points in the AD. It means the first time the AD crosses the real axis is at $kl = 2.61$. Then it crosses again at 5.18, then at 6.86, and so on

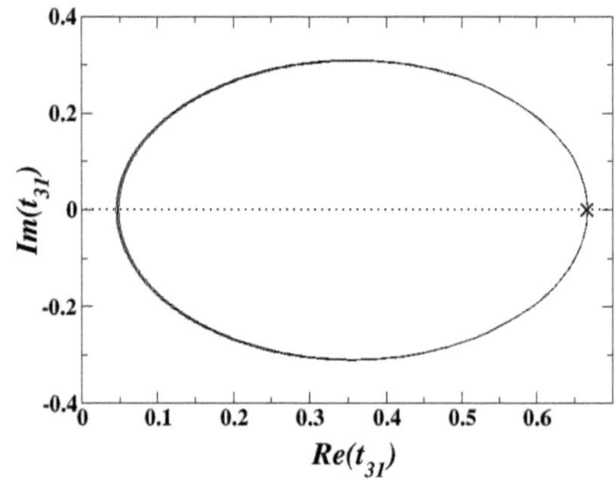

Fig. 6.9 AD for t_{31} for the system as in Fig. 6.7 for the same parameter values as that in Fig. 6.7 but with $l_1 = l_3 = 0$, and $kl = 8.22$. e_0Ul is varied from –1000 to –1050. Note that the starting point is marked with a x which is the same as the point marked 8.22 in Fig. 6.8 as the parameter values corresponding to it are the same

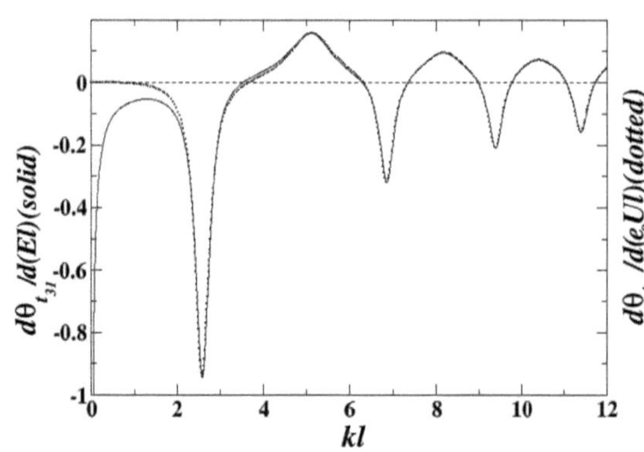

Fig. 6.10 For the same choice of parameter values as in Figs. 6.8 and 6.9. For constant scattering potential in the sample we denote functional derivative as derivative

remain undispersed, satisfying the stationary phase approximation. We have plotted $\theta_{t_{31}}$ versus kl in Fig. 6.11. We can clearly see that $\theta_{t_{31}}$ increases with a point of inflection and for the drops one can clearly see a linear regime. The linear regimes

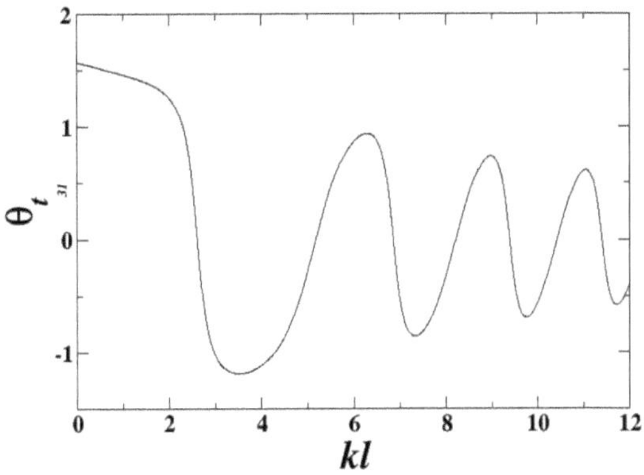

Fig. 6.11 For the same choice of parameter values as in Fig. 6.8

are narrow but one can always construct a wave-packet using the continuum of k values in this linear regime. That wave-packet will be sharply peaked in k space but that is not a problem for detection of the wave-packet as can be inferred from Eq. 6.9. Note that $\theta_{t_{31}}$ (denoted as η_k in Eq. 6.14 and observed comparing Eqs. 6.14 and 6.17), plotted in Fig. 6.11 for the parameters corresponding to Fig. 6.10, can increase as well as decrease. In contrast to that, in case of a 1D barrier or well the scattering phase shift increases monotonously. Effective distance the packet travels is $L + \frac{\eta_k}{k}$ in time $(\tau_0 + \Delta\tau_0)$ and we get all possibilities like $\Delta\tau_0 < 0$ or $\Delta\tau_0 > 0$, there being no momentum operator to deny a possibility (see just after Eq. 6.9). In real space, going back in time is only possible by accelerating beyond the speed of light. In case of Fig. 6.5, by the time we reach $kl = 6.5$ the dispersion due to scattering by the two arms of length $l_1 = l_3 = l$ in Fig. 5.6 becomes negligible and we get agreement between $\frac{d\theta_{t_{31}}(E)}{dE}$ and $\frac{d\theta_{t_{31}}(E)}{d(e_0U)}$. Although they become mostly positive, they do show broad energy ranges where they are identically negative and Eq. 6.23 is satisfied.

Electrons in a wave-packet with negative traversal time can move in negative time and so they will behave like positrons. A positron will attract an electron and annihilate each other to form a new particle or quasi particle. High energy phenomenon of producing photons will not occur in quantum wires as it happens for real positrons. However, this new quasi particle can be quite stable due to the fact that negative times are coming from the solution of the fundamental equations of motion, and so they may move as a bound state. This is an alternate explanation to the numerical calculations presented in Singha Deo [4] and also discussed in Chap. 5. In conclusion, we wanted to demonstrate Eq. 6.23 for certain range of energies and we have successfully shown that in Fig. 6.5 as well as in Fig. 6.10. This also happens for another potential that can be solved exactly in Q1D, that of a negative delta function potential in a multichannel quantum wire. We have checked this for a two channel quantum wire in Chap. 4. Our arguments based on AD and BC suggest that this is a general feature of Fano resonances that lead to the formation of sub-loops in the AD. Although η_k comes from the time-independent Schrodinger equation, but at least in case of completely closed sub-loops in the first Riemann surface, it is the topology of the complex manifold of $Im(s_{\alpha\beta})$ versus $Re(s_{\alpha\beta})$, that primarily determines this η_k as a function of k and not directly the equation of motion. The equation of motion can give different shapes of the sub-loops but that is not important. Physical consequence of Eq. 6.23 is that a signal in the form of a series of wave-packets or a modulated wave can be propagated in negative time. We also conjecture that one can have a bound state of two electrons as a consequence. This chapter is based on our published work [5].

6.5 Reflections

While time travel has been a matter of speculation and contradiction leading to many philosophical questions, we have here by provided a definite mechanism that may carry an information or signal to the past. Such a mechanism was not known so far. One may see [6] for comparison with experimental data. We have shown in the next chapter that some further confirming experiments can be done. There is no contradiction in accepting the fact that a signal can be sent to the past quantum mechanically and has been talked about since the time of Wigner. We have shown that one can construct a Wigner type wave-packet that can carry a signal to the past if we construct a wave-packet with a continuum of k values in a narrow regime of negative phase drops due to Fano resonances. What could be the consequence of such a signal on our life is a matter of immense speculation that even Hollywood fantasizes on. Any philosophical criticism can always be countered by an equally mind boggling philosophy and so in physics (as a discipline) we restrict ourselves to the nuts and bolts of the mechanism involved and this pays off. We did observe the discipline of polishing our shoes before we went to school as kids although within some "-ism" one can always say that going to school barefooted can do us a lot of good.

Einstein's special theory of relativity (STR) is probably a theory to which there are no known violations as long as we do not want to talk about any form of acceleration. All other theories work very well in some regime and fail beyond that and also have theoretical inconsistencies. Similarly, at least for von-Neumann quantum mechanics there are no known inconsistencies. While the standard model or general theory of relativity talks about an incomplete picture, within the low energy regime as long as single particle wave-function modulus gives probability (gauge symmetry), quantum mechanics gives a complete picture and within boosted reference frames (Lorentz's symmetry) STR gives a complete picture. Special theory of relativity reveals the fact that past, present, and future are all equally real and simultaneity is relative. By moving faster and faster we go slower to the future but as we cannot exceed the speed of light we cannot go to the past. The twin cannot go out and return to earth, younger than when he started. Now moving faster in STR is not a physical process achieved through acceleration and in the twin paradox, return to earth has to be without involving deceleration in STR. Brian Cox and Jeff Forshaw in their book “Why does $E = mc^2$?” have done the calculation for such a physical process involving acceleration and deceleration using general theory of relativity (GTR) and the age difference between the twins is way off from what we get from STR. STR reveals just the fact that nature has a symmetry captured by a transformation called boost which is at the heart of this age difference. Now physical acceleration is not very well understood yet and it is not simultaneously consistent with GTR and quantum mechanics. Long ago the Greek philosopher Zeno had argued in terms of deceleration that motion is undefined. He considered a process in which in subsequent time intervals a traveler always goes half of his previous step. Today we analyze his set up using calculus and we understand the problem in terms of converging series (a Taylor expansion is at the root of a converging infinite series and to define the coefficients in a Taylor expansion we need calculus) but that does not mean we have solved the problem of acceleration and defined motion. Today we argue in terms of Unruh effect and Killing vectors and one can emphatically say that motion is undefined. In fact it is related to our poor understanding of the continuum wherein we do not know if there is an answer to the following question. If a bird physically flies from point A to point B we can magnify and start seeing its intermediate steps but is there a limit to how many intermediate steps exist? If there is a limit or a finite bound then the bird is doing some vanishing acts at certain points. Symmetry just captures the geometry of the manifold in which the bird flies and does not care for vanishing acts or continuous motion. This however does not mean that the symmetry captured by boost is to be questioned. That so far is beyond all doubts.

Same can be said about quantum mechanics (up to first quantization). It reveals that nature has a symmetry and that particles like waves have a gauge symmetry. Physical processes have to be understood with reference to this symmetry but there are plenty of physical phenomena (even at low energies) involving transient phenomenon, bistabilities, and shocks that are beyond the axioms of quantum mechanics. But that basic symmetry has never been doubted. In quantum mechanics we can talk of time dependent potentials and we can talk of acceleration in STR but within the framework of symmetries and adiabatic processes.

References

1. Meena K, Singha Deo P (2023) Physica E 149:115680
2. Landauer R, Martin Th (1994) Rev Mod Phys 66:217–228
3. Buttikker M, Landauer R (1982) Phys Rev Lett 49:1739
4. Singha Deo P, Pramana (2002) J Phys 58:195
5. Singha Deo P, Satpathi U (2019) Results Phys 12:1506
6. Satpathi U, Singha Deo P (2016) Ann Phys 375:491

Chapter 7
Mesoscopic Response

Abstract In this chapter we discuss various experiments that can be done to observe and confirm negative LPDOS and hence time travel. All these experiments can be done with present day technologies. We also talk about theoretical experiments that reveal these facts.

In earlier chapters we have discussed that certain experiments have already been done and one has to re-interpret the data to support our results. In Ref. [1] and in Chap. 5, we have argued that the experimental data has to be also represented in the form of Argand Diagrams (AD) so that we can see the loops and sub-loops, rather than plotting scattering phase shifts and cross sections separately. We can see in Fig. 5.3c and d that a small change in scattering cross section is associated with a large change (in quality and quantity) in phase shifts and vice verse. Data can be represented in multiple ways but the ADs allow us to interpret in terms of topology and analysis which goes beyond quantum mechanics. We believe there are many other subtle aspects in ADs that are very informative and one has to study them in the future. To do a proper justice we need more densely obtained experimental data plotted in the form of ADs. Below we talk of some other experiments that can be done with present day technology that can tell us more about the hierarchy of DOS, mesoscopic response, and time travel. This kind of experiment has already been done but one has to redo them focusing on specific aspects to be highlighted below.

7.1 Three Probe Set Up: A Theoretical Experiment

Here we consider a special case of the system shown in Fig. 3.2 in the sense that we have only two fixed leads and an STM tip making it a three probe set up [2]. The STM tip is given a lead index β and the fixed leads are indexed γ and α as shown in Fig. 7.1. The STM tip is such that we can vary its proximity to the sample and thus can have four possible functions. First case is that it does not make an actual contact and also does not draw or deliver any current but can locally change the electrostatic potential by $\delta U(\mathbf{r})$ at a point $\mathbf{r}$. Second case is that it does not make a contact but can

P. S. Deo and K. Meena, *Time Travel: A Reality in Mesoscopic Physics*,
https://doi.org/10.1007/978-981-96-1136-2_7

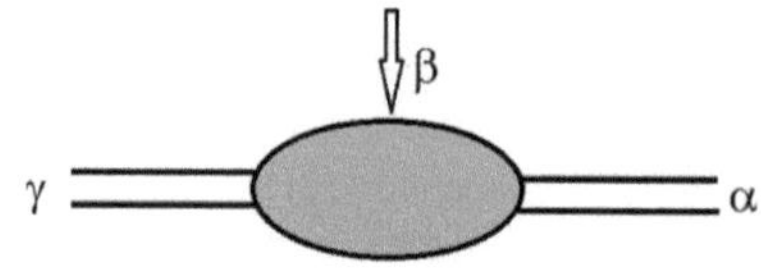

Fig. 7.1 Here we show a simpler version of Fig. 3.2 as now there are only two fixed leads indexed γ and α apart from the STM tip β. This is a cartoon of a realistic experimental set up that helps us measure some of the lower members in the hierarchy of DOS

draw or deliver a current via weak coupling to the states of the system. What it means is that the STM tip is weakly coupled to the states of the sample and exchange of current does not alter the states of the sample to an approximation. Third case is that it makes an actual contact and becomes like any other lead. Fourth is that it makes a contact and yet does not draw or deliver a current because its chemical potential is so adjusted. Fourth case is the typical situation of the Landauer-Buttiker three probe conductance set up which is now well established as a mesoscopic phenomenon, greatly studied theoretically as well as experimentally [2]. In this fourth case, leads γ and α are current leads and since the lead β does not draw or deliver any current, it is a voltage probe that gives the voltage of the floating point $\mathbf{r}$ to which the STM tip is attached. We will first analyze the second case with respect to negative DOS members and show that it is a paradigm to indirectly test time travel experimentally. And then we will also show how these predictions naturally come up also in the well studied case four and not noticed before as the specific design was not considered.

We will first show how some of the lower members of DOS can manifest in this experimental set up of Fig. 7.1 that can be measured. Here the sample is the shaded region between the two leads that are indexed γ and α while the STM tip is indexed β. We have seen in Chap. 4 that such a realistic system can generally exhibit Fano resonances. At zero temperature below the Fermi energy the transmission probability T of quantum mechanics is $\frac{h}{e_0}$ of electronic current (see Eq. 1.9) at that particular energy E and that is how the following formulas are to be interpreted. Let us consider the situation when the tip of β is not making a physical contact with the sample but close enough to deliver (or draw) a current to (or from) the sample due to weak coupling (case II). Experimentally we can create special situations when only one lead injects current and all other leads carry current away or only one lead emits current while all other leads inject current. Otherwise, current could be coming from any lead and taken away by any lead and as said before it is not possible to trace which current goes where, but the following quantities that depend on members of DOS hierarchy can experimentally manifest [3].

$$T^e_{\beta\alpha} = 4\pi^2 \nu_\beta |t|^2 \rho_i(E, \mathbf{r}, \alpha) \tag{7.1}$$

$$T^i_{\alpha\beta} = 4\pi^2 \nu_\beta |t|^2 \rho_e(E, \alpha, \mathbf{r}) \tag{7.2}$$

$$T^e_{\beta\gamma} = 4\pi^2 \nu_\beta |t|^2 \rho_i(E, \mathbf{r}, \gamma) \tag{7.3}$$

$$T^i_{\gamma\beta} = 4\pi^2 \nu_\beta |t|^2 \rho_e(E, \gamma, \mathbf{r}) \tag{7.4}$$

Here ν_β is the density of states in the lead β that couples to the states at the point **r** through the coupling parameter t. Equations 7.1 and 7.3 correspond to current drawn by the lead β while Eqs. 7.2 and 7.4 correspond to that delivered by lead β. For example, transmission probability $T^e_{\beta\alpha}$ is contribution of lead α to emission current taking place through lead β and it is proportional to the injectivity of lead α to the remote point **r** where STM tip β is situated. Others can be similarly interpreted.

Let us now set up the biases in Fig. 7.1 correspond to the situation such that classically electrons can be incident along only one lead and that is γ and scattered to α or β. Lead β is weakly coupled to the sample (shaded region in Fig. 7.1) in order to correspond to case 2 described above. Let chemical potential of left reservoir connected to lead γ be μ_γ, and chemical potential of right reservoir connected to lead α be μ_α, and the other end of the lead β is earthed. For incident energy E such that $0 = \mu_\alpha < E < \mu_\gamma$, only lead γ should inject a current according to our classical notions. Such a set up can be obtained in the laboratory and we want to address the coherent current flowing from γ to α. It was shown [3] that in such a situation

$$|s'_{\alpha\gamma}|^2 - |s_{\alpha\gamma}|^2 = -4\pi^2 \mid t \mid^2 \nu_\beta \rho_{lpd}(E, \alpha, \mathbf{r}, \gamma) \tag{7.5}$$

where $s'_{\alpha\gamma}$ is the scattering matrix element for scattering from γ to α when the STM tip is drawing a current given by the term on RHS. When the STM tip is removed by $t \to 0$ then this scattering amplitude will be $s'_{\alpha\gamma} = s_{\alpha\gamma}$. Intuitively speaking, ρ_{lpd} is positive and the RHS is negative accounting for loss of coherent electrons to the earthed lead β.

Scattering probabilities $|s'_{\alpha\gamma}|^2$ and $|s_{\alpha\gamma}|^2$ are well defined in quantum mechanics and when multiplied by a factor $\frac{h}{e_0}$ they give the measured mesoscopic current from γ to α. One of the facts that has emerged from the study of Fano resonances in this book is that $\rho_{lpd}(E, \alpha, \mathbf{r}, \gamma)$ can be positive as well as negative, implying that $|s'_{\alpha\gamma}|^2 - |s_{\alpha\gamma}|^2$ can be also positive as well as negative according to Eq. 7.5. When negative then the sample will lose coherent current to the earthed STM tip, and when positive it will draw coherent current from the earth which is counter-intuitive. As $|s'_{\alpha\gamma}|^2$ and $|s_{\alpha\gamma}|^2$ are both measurable and so if the equality in Eq. 7.5 can be justified, then in relative proportions $\rho_{lpd}(E, \alpha, \mathbf{r}, \gamma)$ is also measurable which is good enough to confirm its sign. One may always try to justify that from a brute force practical experiment, but we will show below that it can be justified by what we call a theoretical experiment. For our theoretical experiment let us consider the system shown in Fig. 7.2. First of all it is a system for which exact quantum mechanical calculations can be made and it is a system that exhibits pronounced Fano resonances. And secondly as pointed out in Chap. 5, there are 6 universality classes based on AD. Note that in the second paragraph of Chap. 5 and in Sect. 5.1 we have said how the topology of the complex plane of $Im(s_{\alpha\beta})$ versus $Re(s_{\alpha\beta})$ gives us these classes of Argand diagrams. For some of those classes there may be no examples but a couple of them can be easily realized and happen because of the nature of resonances involved. For

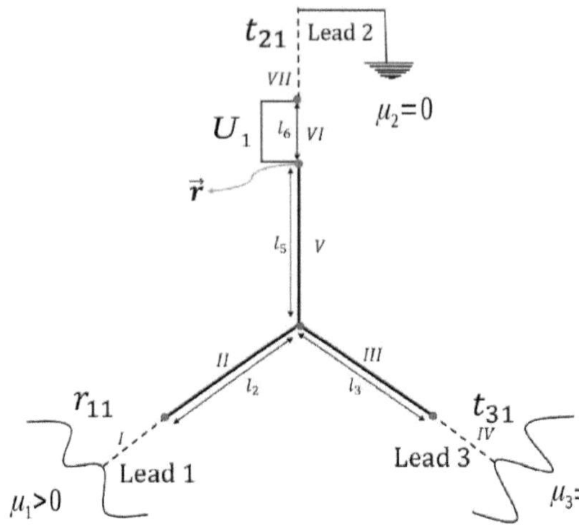

Fig. 7.2 The sample is the three prong potential of Fig. 5.6 shown by the solid lines and the entire system consists of the sample connected to three reservoirs via three leads. Different regions of the system are marked by Roman numbers I, IV, and VII being the leads, shown by dashed lines. Lead 2 is made exactly like that in Fig. 5.6 by earthing it and the chemical potential of lead 3 is also set to zero. In that case for a small positive chemical potential at lead 1 will create a situation wherein only lead 1 injects a current while the two other leads carry some current away from the sample. Lead 2 connects to the sample through a tunneling barrier shown as region VI. Lengths of different regions are shown as l_2, l_3, etc. This system is a 1D version of the system in Fig. 7.1 where lead γ is renamed as lead 1, etc., and as a result $t_{31} \equiv s_{31}$, $t_{21} \equiv s_{21}$, and $r_{11} \equiv s_{11}$

example Breit-Wigner resonance gives AD that is not closed in one Riemann surface but goes round the singularity. We know the system in Fig. 7.2 will produce an AD that has closed sub-loops within a single Riemann surface that does not enclose the phase singularity as a consequence of Fano resonance. Since it is a matter of finding a particular class of AD, one can study a 1D system as a representative of a class. In this sense the system in Fig. 7.1 at Fano resonances with the modification that the upper lead is weakly coupled to earth via a strong potential can be represented by the 1D system in Fig. 7.2. Besides if it is a question of whether LPDOS is present at all in a quantum system and more importantly whether its negativity is allowed in quantum mechanics, then demonstrations in 1D quantum mechanics should suffice.

One has to further observe the fact that quantum mechanics is a theory for objects to be found in nature but everything that can be calculated from quantum mechanics may not be found in nature. Quantum mechanics does not prohibit the possibility of calculating an object like $\rho_{lpd} = -\frac{1}{2\pi} |s'_{\alpha\gamma}|^2 \frac{\delta\theta_{s'_{\alpha\gamma}}}{\delta U(\mathbf{r})}$, or say any higher order derivative, but cannot ascertain its occurrence in nature. Such a coordinate dependent LPDOS cannot be defined from the wave-function at the point $\mathbf{r}$ and so cannot be reconciled with the probabilistic interpretation of quantum mechanics. In quantum mechanics an electron coming from γ can go to the STM tip or to the lead α or can get reflected back rather randomly and there is no equation of motion for such an electron. Schrodinger equation is an equation for an ensemble of electrons and gives a probabilistic answer of $T_{\alpha\gamma} = |s_{\alpha\gamma}|^2$ completely ignoring how the individual electrons are behaving. So within the axiomatic framework of quantum mechanics one cannot provide a quantitative or qualitative value for the number of γ to α bound electrons that pass through the point **r** of the sample and this has been discussed in detail in Chap. 3 while discussing Eqs. 3.56 and 3.57. In short, the LHS of Eq. 7.5 is physical according to quantum mechanics but the RHS is not. One can see from quantum mechanics that for the system in Fig. 7.2

$$|s'_{31}|^2 + |s'_{21}|^2 + |s'_{11}|^2 = 1$$

$$|s_{31}|^2 + |s_{21}|^2 + |s_{11}|^2 = 1$$

where $|s_{31}|^2 = |t_{31}|^2$, etc. The primed and unprimed quantities may correspond to different parameter values of the system which is a consequence of unitarity or current conservation. As a special situation for the unprimed case, lead 2 can be removed with the help of a high tunneling barrier which will mean $|s_{21}|^2 = 0$ but for the logic below we retain this term. Taking a difference we get

$$|s'_{31}|^2 - |s_{31}|^2 = |s_{21}|^2 - |s'_{21}|^2 + |s_{11}|^2 - |s'_{11}|^2$$

which means the LHS of Eq. 7.5 is well defined in quantum mechanics.

Larmor clock on the other hand uses analyticity of scattering matrix elements and gives ρ_{lpd} as a physical quantity that says the number of electrons going from γ to α access these partial states that are located at the point **r**. Thus Eq. 7.5 given by Buttiker says that in a quantum system, there is a local coordinate dependent DOS called LPDOS that can manifest in experiments although the standard theory of quantum mechanics says that is not possible. Existence of LPDOS failed to pass the test for two reasons.

1. The original derivation of ρ_{lpd} or LPDOS in Chap. 3 uses the analyticity property of scattering matrix element but also has to use a number of ingredients from quantum mechanics as well as classical mechanics, for example spin precession and Larmor frequency.
2. Also there was no way to reconcile the numerical values of LPDOS with WDT in general or in regimes when both of them become negative.

The issue of negativity was explained in Chap. 6 as a signature of time travel that manifests in complete consistency between WDT and LPDOS if one makes use of special resonances called Fano resonances which are easy to find in mesoscopic systems as shown in Chap. 4. Time travel experiments may still be difficult to conduct practically. While the kind of experiments schematically seen in Figs. 7.1 and 7.2 has been already done in mesoscopic systems. However, without relying on a brute force experiment we will discuss a theoretical experiment, in the following sense.

1. We will establish the equality in Eq. 7.5, purely from analyticity, without referring to spin precession or any input from any mechanics, but purely from the topology

of the complex plane for $Im[s_{\alpha\beta}]$ versus $Re[s_{\alpha\beta}]$. One may think that we need Schrodinger equation to calculate $s_{\alpha\beta}$, but in fact topology around a singularity has some universal features as discussed in the paragraph after Eq. 7.5. There are in fact many examples wherein one finds that close to a singularity the equations of motion like Navier-Stokes equation or the equation of state for a gas become secondary and the singularity starts determining robust phenomena. One may also cite that craftsmen involved in making hand made knives have discovered through evolution that the sharpness of the edge of a knife depends on the grind all the way up to the spine of the knife. For example a sabre grind is very sharp while a scandi grind is more durable. In this case the edge of the knife is a singularity, while the grind is classification of the surface of the knife away from the singularity. In this sense we can truly bypass the axioms of quantum mechanics to establish the equality in Eq. 7.5.

2. We will not only show that sign changes of LHS are related to sign changes of RHS but there can also be quantitative agreement implying the fact that the RHS of Eq. 7.5 is also physical.

We will essentially use AD and BC at Fano resonances. Since we have already established that the same principles can lead to time travel, one can interpret that for this set up, drawing a coherent current from the earth is essentially losing coherent electrons back in time.

We know that quantum states on an infinite 1D line are given by Schrodinger equation with DOS being $\frac{1}{hv}$ independent of whether these states are occupied by Bosons or Fermions and independent of temperature. Similarly we say that the STM tip has a DOS given by ν_β and the point $\mathbf{r}$ in the sample with which the STM tip interacts has an over specified DOS or a DOS hierarchy member. Thus there are two DOS that feature in these formulas in Eqs. 7.1–7.5 that are ν_β and an $\mathbf{r}$ dependent DOS. There is a factor $\frac{1}{2\pi}$ for each of them that cancels the factor $4\pi^2$. This is the $\frac{1}{2\pi}$ factor that appears in Eq. 3.49. There can be a very simple interpretation for these factors as follows. LHS of Eq. 3.49 is a DOS that can accommodate electrons that are countable but $2\pi\rho(E,\gamma)$ is uncountable, the RHS of Eq. 3.49 being in terms of $\theta_{s_{\alpha\gamma}}(E)$ corresponding to a continuous rotation. This is also why we get a factor 2π on the RHS of Eq. 5.1.

Let us drop the factor $|t|^2\nu_\beta$ from Eq. 7.5 because such factors will be only needed if we are using a commercially available STM tip, where this factor characterizes the STM tip. This requires us to drop a 2π factor in Eq. 7.5 for reasons mentioned in the previous paragraph. For the case of Fig. 7.2, this means for a first principle calculation for the system in Fig. 7.2 we expect (if Eq. 7.5 is correct) that

$$|s'_{\alpha\gamma}|^2 - |s_{\alpha\gamma}|^2 = -2\pi\rho_{lpd}(E,\alpha,\mathbf{r},\gamma) \tag{7.6}$$

Now to make it correspond to Fig. 7.2

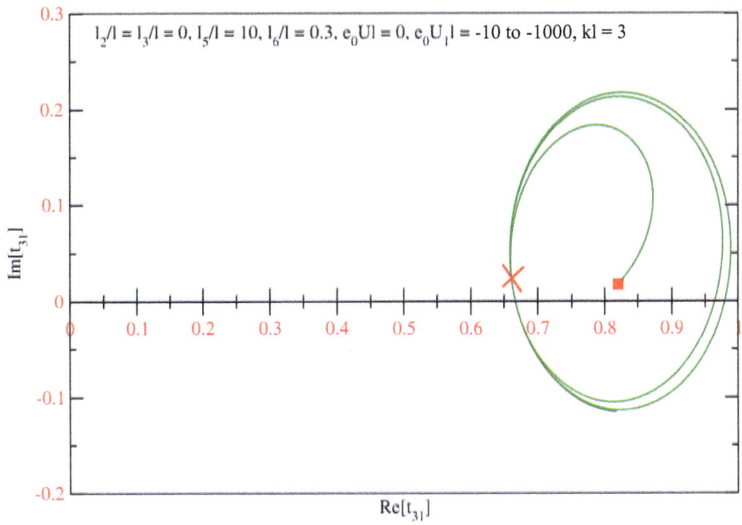

Fig. 7.3 In this figure we plot the AD for t_{31} of the system shown in Fig. 7.2 for the coupling potential U_1 varying in a range that gives three sub-loops, all within one Riemann surface. The starting point is marked by a small square block corresponds to $e_0U_1l = -10$. The end point is unmarked and corresponds to a value $e_0U_1l = -1000$. All the sub-loops smoothly come back to a point marked by a cross. Other parameters are mentioned inside the figure

$$|t'_{31}|^2 - |t_{31}|^2 = -2\pi\rho_{lpd}(E, 3, \mathbf{r}, 1) \tag{7.7}$$

In Fig. 7.3 we plot the AD for t_{31} for the system shown in Fig. 7.2 by varying the potential U_1 which connects the three prong potential with lead 2 that is earthed. The AD diagram makes smooth sub-loops within one Riemann surface and each sub-loop is due to a Fano resonance. Any physical quantity that depends on t_{31} will go through a cycle over one particular sub-loop. For one closed sub-loop generated by monotonously varying a parameter, say U_1

$$\oint_c \Delta\theta_{t_{31}} = 0$$

where $\theta_{t_{31}} = Arctan\frac{Im[t_{31}]}{Re[t_{31}]}$

$$\text{or} \quad \oint_c \frac{\delta\theta_{t_{31}}}{\delta U_1(\mathbf{r})}\,\Delta U_1(\mathbf{r}) = 0$$

$U_1(\mathbf{r})$ being a local potential at the fixed point $\mathbf{r}$.

$$\text{or} \quad \oint_c \frac{1}{2\pi}|t_{31}|^2\,\frac{\delta\theta_{t_{31}}}{\delta U_1(\mathbf{r})}\,\Delta U_1(\mathbf{r}) = 0 \tag{7.8}$$

$$\text{or} \quad -\oint_c \rho_{lpd}(E, 3, \mathbf{r}, 1)\, \Delta U_1(\mathbf{r}) = 0 \tag{7.9}$$

Similarly over the same sub-loop,

$$\oint_c \frac{\delta |t_{31}|^2}{\delta U_1(\mathbf{r})} \Delta U_1(\mathbf{r}) = 0$$

$$\oint_c \Delta |t_{31}|^2 = 0 \tag{7.10}$$

Essentially

$$\oint_c \Delta |Re[t_{31}] + Im[t_{31}]|^2 = 0$$

Therefore from Eqs. 7.8 and 7.10 we get the equality

$$\oint_c \Delta |t_{31}|^2 = \oint_c \frac{1}{2\pi} |t_{31}|^2 \frac{\delta \theta_{t_{31}}}{\delta U_1(\mathbf{r})} \Delta U_1(\mathbf{r}) = 0 \tag{7.11}$$

purely as a consequence of the topology of a complex plane. The topology also tells us that for loops that are not completely closed in one Riemann surface like that in the case of BW resonance, there will be correction terms. If we want to compare the integrands in Eq. 7.11, then we need an extra factor of 2π as discussed before. Thus

$$\Delta |t_{31}|^2 \approx 2\pi \frac{1}{2\pi} |t_{31}|^2 \frac{\delta \theta_{t_{31}}}{\delta U_1(\mathbf{r})} \Delta U_1(\mathbf{r}) \tag{7.12}$$

Viewing a derivative as an effect of an infinitesimal change in U_1 we get

$$|t'_{31}|^2 - |t_{31}|^2 \approx |t'_{31}|^2 (\theta_{t'_{31}} - \theta_{t_{31}}) \tag{7.13}$$

where primed quantities and unprimed quantities are calculated for an infinitesimal difference of U_1. If we want to compare integrands then we can only write an approximate equality because unlike in the case of Eq. 6.26 where we were comparing the same object over two sub-loops that are counterparts of each other, here we are comparing two different objects over one sub-loop. There will always be in the least a phase difference between the two quantities.

Also from Eqs. 7.9 and 7.10 we get

$$\oint_c \Delta |t_{31}|^2 = -\oint_c \rho_{lpd}(E, 3, \mathbf{r}, 1)\, \Delta U_1(\mathbf{r}) = 0 \tag{7.14}$$

Therefore, it is obvious that $\Delta |t_{31}|^2$ will go through a positive-negative cycle over a closed sub-loop as a consequence of the topology of the relevant complex plane. Given the fact that LHS is physical, the cycles of ρ_{lpd} are also physical as they are just

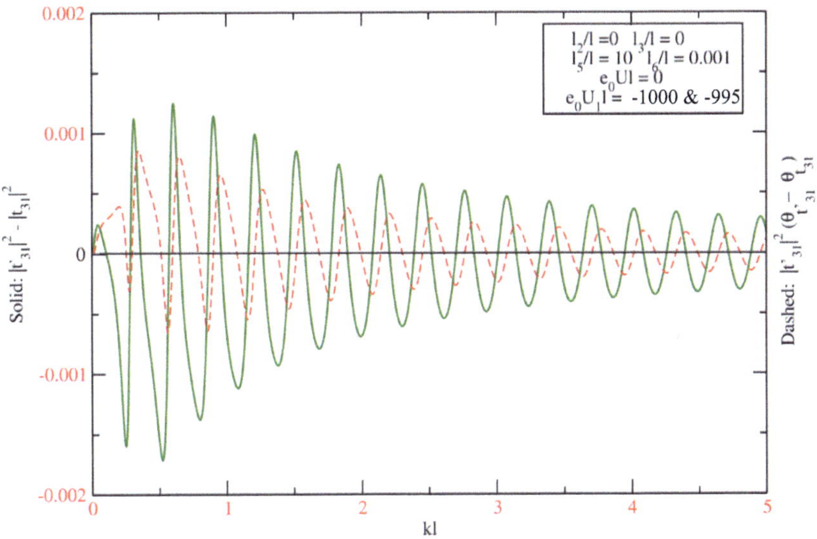

Fig. 7.4 In this figure we are plotting the LHS and RHS of Eq. 7.13 to show that they both can oscillate between positive and negative values. The primed and unprimed values are for small differences in U_1 at a k value that we vary continuously. The sign change and magnitude of both the curves originate from the smooth cyclic AD of Fig. 7.3

different expressions of the same sub-loop made by the AD of t_{31}. Fano resonances can be used to give a clear indication of that although the principle may work for other resonances.

In Fig. 7.4 we plot LHS and RHS of Eq. 7.13 where the primed values and unprimed values are again for small differences in U_1 as the incident wave vector k is varied and that is the horizontal axis. It becomes clear that LHS and RHS of Eq. 7.13 oscillate with positive and negative values. Initially the two curves are a little different but as k increases, effects of dispersion are minimized, and magnitudes of the two curves are similar with a phase difference between them. The oscillations of both the curves are due to the smooth cyclic nature of the AD in Fig 7.3. In other words, the underlying principle behind the oscillation of both curves is the same. The two curves oscillate as a result of quantum interference. But the symmetric oscillation between positive and negative values and the similarity between the two curves can only be noted if there are well defined Fano resonances and signify the fact that the RHS of Eq. 7.5 is physical. For BW resonances the last two features are not seen.

We are all along considering the temperature to be zero wherein the number of states is equal to the number of electrons and in that case it is easy to switch between statistical mechanics at zero temperature and pure quantum behavior. Now in statistical mechanics we can change many parameters and study an observable current. In the subsequent discussion, we will describe a set up that requires us to tune and adjust the chemical potential μ_β. We focus on the situation shown in Fig. 7.5 where we have also taken an applied Aharonov-Bohm flux as that helps separate coherent

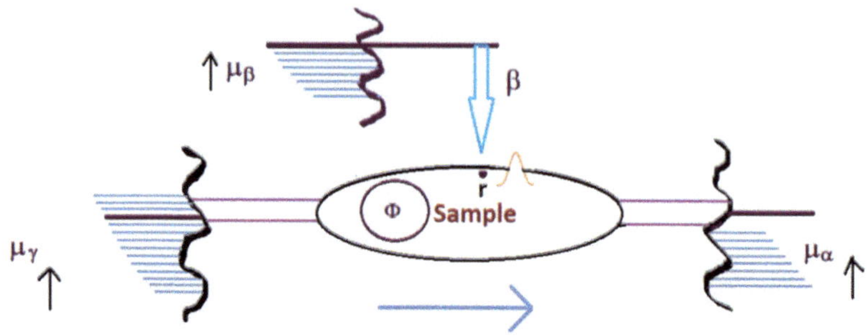

Fig. 7.5 The role of lead β (an STM tip) is to inject or remove electrons from the system as well as causing some decoherence. The STM tip makes an actual contact with the sample at point **r** and μ_β is so adjusted that there is no net current flowing through the STM tip but it can still cause decoherence. At 0°K, a current flows from lead γ to lead α in the energy interval $(\mu_\gamma - \mu_\alpha)$. One can put an Aharonov-Bohm flux Φ through the sample which is very useful to separate coherent and incoherent effects

and incoherent processes. Here the reservoirs are explicitly shown as electron reservoirs with definite chemical potentials μ_γ, μ_α, and μ_β. In a situation wherein the probe β makes an actual contact with the sample we get a three probe set up and also the probe β is made like a voltage probe in the sense that its chemical potential is so adjusted that it does not draw any net current from or into the system. This set up is ideal for studying the celebrated Landauer-Buttiker three probe conductance given by [3]

$$G = -G_{\alpha\gamma} - \frac{G_{\alpha\beta}G_{\beta\gamma}}{G_{\beta\alpha} + G_{\beta\gamma}} \tag{7.15}$$

Here

$$G_{\alpha\beta} = \frac{e_0^2}{h}|s_{\alpha\beta}|^2 \quad for \quad \alpha \neq \beta \quad etc. \tag{7.16}$$

is the β to α conductance. This formula can be rewritten in terms of the hierarchy of the density of states in the following way [3].

$$\begin{aligned} G = \frac{e_0^2}{h}(|s_{\alpha\gamma}|^2 - 4\pi^2|t|^2\rho_{lpd}(\alpha, \mathbf{r}, \gamma) \\ +4\pi^2|t|^2\frac{\rho_e(\alpha, \mathbf{r})\rho_i(\mathbf{r}, \gamma)}{\rho_{ld}(\mathbf{r})}) \end{aligned} \tag{7.17}$$

So the three terminal formula of Eq. (7.15) is now restated in the form of Eq. (7.17). We will try to understand Eq. 7.17 when ρ_{lpd} is positive as that will naturally tell us what will happen if it is designed to be negative. Note that in the above formula if the lead β is completely removed then $|t|^2 = 0$ and we will be left with only the first of the three terms. This is the standard two probe Landauer conductance formula. The second term comes with a negative sign and accounts for the loss of coherent electrons due to the lead β as can be concluded by comparing with the RHS of Eq. 7.5. This comparison tells us that when lead β makes an actual contact then ν_β has no role and so in the three probe Landauer conductance formula, only the DOS of the incoming lead matters. These lost coherent electrons would have escaped to the Earth in case of Fig. 7.2 but now they are not escaping as β is not drawing any net current. Only those partial electrons that are going from γ to α coherently are targeted in both the cases and hence this term depend on Aharonov-Bohm flux in the same way as $|s_{\alpha\gamma}|^2$, reducing the overall flux dependence of G in Eq. 7.17. Note that Eq. 7.17 also suggests that while quantum mechanics is non-local, decoherence happens locally. Meaning an STM tip attached to point $\mathbf{r}$ gives the last two terms on the RHS of Eq. 7.17 that does not depend on any other point $\mathbf{r}'$.

The lost electrons are momentarily incoherent particles at the point $\mathbf{r}$ and eventually redistribute to γ and α. The question arises what will be the ratio of this redistribution and this will again be determined by the members of the hierarchy. Redistribution contribution to G is the third term in Eq. 7.17 separately written below.

$$4\pi^2|t|^2\frac{\rho_e(\alpha,\mathbf{r})\rho_i(\mathbf{r},\gamma)}{\rho_{ld}(\mathbf{r})} \tag{7.18}$$

Note that this term consists of the product of two separate probabilities associated with two independent processes. Each of the two separate processes is however coherent, that is one coherent process gives a $\rho_e(\alpha,\mathbf{r})$ and the other gives $\rho_i(\mathbf{r},\gamma)$. One is an emissivity from $\mathbf{r}$ to α and the other an injectivity from γ to $\mathbf{r}$. Essentially a coherent current loss from γ to α via $\mathbf{r}$ is compensated by a gain due to two coherent terms, one from γ to $\mathbf{r}$ and the other from $\mathbf{r}$ to α, due to decoherence at the point $\mathbf{r}$. Hence the gain term and the loss term will be of opposite signs.

To understand the denominator in Eq. 7.18 let us consider the following. Total number of incoherent electrons at the point $\mathbf{r}$ must be

$$4\pi^2|t|^2\frac{\rho_e(\alpha,\mathbf{r})\rho_i(\mathbf{r},\gamma)}{\rho_{ld}(\mathbf{r})} + 4\pi^2|t|^2\frac{\rho_e(\gamma,\mathbf{r})\rho_i(\mathbf{r},\gamma)}{\rho_{ld}(\mathbf{r})} \tag{7.19}$$

$$= 4\pi^2|t|^2\frac{\Big(\rho_e(\alpha,\mathbf{r})+\rho_e(\gamma,\mathbf{r})\Big)\rho_i(\mathbf{r},\gamma)}{\rho_{ld}(\mathbf{r})} \tag{7.20}$$

The first term in Eq. 7.19 is just the term in Eq. 7.18 and gives the fraction of incoherent electrons at $\mathbf{r}$ that goes to α and the second term is that which goes to γ, which means Eq. 7.20 gives the total incoherent electrons at the point $\mathbf{r}$ originating due to the loss of γ to α coherent electrons contributing to the second term on the RHS of Eq. 7.17. Therefore, in a situation when lead β is not drawing or delivering current, the quantity in Eq. 7.20 will be equal and opposite to the second term on the RHS of Eq. 7.17. This however is not a mathematical equality but an equality due to the balancing act of μ_β. The second term in Eq. 7.17 is independent of μ_β and the balancing act of μ_β makes Eq. 7.20 equal to it. Given the fact that $\rho_e(\alpha,\mathbf{r})+\rho_e(\gamma,\mathbf{r})=\rho_{ld}(\mathbf{r})$,

Eq. 7.20 is simply proportional to $\rho_i(\mathbf{r}, \gamma)$. Essentially the emissivity of the point $\mathbf{r}$ to all possible outgoing channels is the local density of states at the point $\mathbf{r}$. This $\rho_i(\mathbf{r}, \gamma)$ is the quantity that has to be balanced against the chemical potential of the lead β at all flux so that lead β does not draw or deliver any net current. This is a situation wherein we are at 0^oK and in the regime of incident energy E being such that $\mu_\gamma > E > \mu_\alpha$. In this regime there is no injectivity from lead α and so only the injectivity of lead γ matters.

So if ρ_{lpd} is made negative then the RHS in Eq. 7.5 becomes positive implying the system draws in coherent electrons to the point $\mathbf{r}$ instead of losing them. This can be also interpreted as losing coherent electrons in reverse time. Thus in that case, the lead β can enhance coherence. The same signature can also be seen from Eq. 7.17 which too can be experimentally verified. Also if the second term on the RHS in Eq. 7.17 becomes positive then $\rho_i(\mathbf{r}, \gamma)$ becomes negative which can be verified by the way one has to balance μ_β. A negative number of states accommodating negatively charged electrons can behave as a positive charge cloud. If it can attract one electron, it can also attract another electron and thus mediate an electron-electron attraction. We have thereby alternately justified Eqs. 7.5 and 7.17 purely from analyticity and topology of complex plane. Quantum mechanics is an axiomatic framework or a model that is artificially tuned to explain a variety of physical phenomenon. We can say from Goedel's incompleteness theorem that there can be some more facts and phenomenon that can be physically seen in quantum systems but never pop up in the axiomatic framework. Therefore it is not a surprise that ρ_{lpd} cannot be explained by quantum mechanics but is very much present in a quantum system. The reason why asymptotic solutions of the Schrodinger equation can be used to calculate ρ_{lpd} is because after all the Hilbert space is isomorphic to the complex plane and has to respect the topology of the complex plane. Within the model, we can talk about precession of a spin or a small time-dependent potential on top of the static potential $U(\mathbf{r})$ [6], the outcome of something that we can calculate from the model has to respect the topology of the complex plane. A model that explains phenomena in a manifold can at no point has the freedom to get out of the manifold. As a crude analogy, some animals have short necks while a giraffe can have a three meter long neck but that is the maximum allowed by the manifold and its curvature. If the giraffe wants a neck longer than that, it has to first get out of the manifold. Within the manifold, it can try various different evolutionary paths but will eventually end up having a three meter long neck. We have also established that the equality in Eqs. 7.5 and 7.17 will be approximate equalities but does signify the fact that sign changes of the terms are determined by the same topological principle that gives an exact equality in case of Eq. 6.23. We will show in the next section that a few more exact equalities follow from the same principles and stated in Eqs. 7.29 and 7.31. All these equalities stem from the fact that we can have topologically closed trajectories (independent of shape) in one Riemann surface of the complex plane $Im(s_{\alpha\beta})$ versus $Re(s_{\alpha\beta})$. Real systems can show such trajectories very generally, but may also show trajectories that are not completely closed within one Riemann surface, but the principle is rather novel and hereby established. That ρ_{lpd} plays the role of I in Eq. 5.1. The results of this section have been published in [4].

7.2 Persistent Currents: A Theoretical Experiment

Evanescent modes are one of the puzzling features of quantum mechanics wherein a particle with energy less than the barrier height can tunnel through the barrier and this has no classical counterpart. Although mathematical analogies can be found with electromagnetic wave-packets, physical interpretations pose a serious challenge. It is not possible to draw any classical correspondence because one cannot construct a wave-packet with such evanescent states and so one cannot define propagation of the particle under the barrier. This has lead to study several aspects of states under the barrier and one such aspect is tunneling time. Since without a wave-packet one cannot define a group velocity and therefore addressing the problem leads to several puzzles [5]. In a situation where a proper theory or formalism is missing making sense of experiments is difficult [6]. One can address the problem with the help of physical clocks like Larmor clock and Wigner delay time but that invariably appeals to semi-classical concepts like spin precision and stationary phase approximations. Feynman path approach has also been applied [6] and different approaches give different results. Hartman effect like phenomenon [7] shows that a tunneling particle can come out of the barrier before entering it which raises questions like if at all there is propagation under the barrier. We describe here a situation wherein one can make a theoretical experiment to confirm current due to evanescent modes and hence conclude propagation under the barrier. That will also be a testing ground for the validity of physical clock like the Larmor clock. Of course one way to validate Larmor clock is to appeal to Burgers circuit and analyticity, which we have done. This will be another case of making an alternate validation by showing that current under the barrier as a measurable quantity can be established from analyticity of scattering matrix elements even if one cannot prove its existence from quantum mechanics alone. We will discuss with respect to the set up described in the next few paragraphs. For the complexity of physical interpretations we limit ourselves to the simplest system of 1D rings. The sample in the form of a 1D quantum ring is not a loss of generality as explained before. Such 1D rings can be achieved experimentally using lateral confinement and materials for which effective mass approximation works [2]. Besides, as we stated in first chapter, one can make finite thickness rings that consist of many independent 1D channels. A finite thickness lead or ring can have multiple channels but DOS per channel is just the 1D DOS that can be achieved in an experiment by sub-band quantization [10–12, 14, 15].

We will first provide a simple description of the mesoscopic set up (see Fig. 7.6) and then discuss how this set up helps us address the problems discussed in the previous paragraph. The mesoscopic sample is taken to be a ring pierced by a magnetic flux Φ through the center of the ring such that there is no magnetic field on the electrons confined to the ring. The ring is usually made up of gold or copper or semiconductors that can accommodate an electron gas. We are interested in the equilibrium response of this sample which in this case is response to a magnetic field. There is a reservoir shown as a 3D block which is a source of electrons at chemical potential μ and temperature T, making it a grand canonical system. The reservoir injects electrons to the sample or to the ring through a lead shown as region A. The ring can thus exchange electrons with the reservoir through the lead. This exchange does not result in a net current in the lead because electrons that go into the ring also come out of it and escape to the classical reservoir which is very typical of a voltage probe. The system therefore constitutes an equilibrium system where the reservoir also acts as a source of decoherence according to the Landauer-Buttiker formalism.

The electrons in the lead and inside the ring are purely described by quantum mechanics. If the ring is isolated then it will have some eigen-functions and eigen-energies that were obtained from Schrodinger equation and shown in Fig. 2.3. But once connected to the reservoir, the states in the ring will be affected. They will no longer be eigen-states of the Hamiltonian and the eigen-energies will acquire some broadening due to life time related effects as the states can now leak into the reservoir. One can use a set up (see Fig. 1.17) to control the broadening and recover the results obtained from the spectrum shown in Fig. 2.3. The states in the lead are unaffected by the ring as in Landauer-Buttiker formalism it is an ideal 1D system with a typical DOS given by $\frac{1}{hv}$. A part of the ring (region C in Fig. 7.6) has a potential U that is shown as a thickened line. U can be so adjusted that an electron can tunnel through this region C. This region C can be extended to the entire ring wherein the entire ring can be made into a tunneling region. We know the AB flux can drive a persistent current in the ring and this current can thus be a tunneling current in part of the ring or in the entire ring. This will allow us to extend the theoretical consistency of quantum currents to the tunneling regime as follows. When the barrier is located only in a part of the ring then the persistent current in the ring can flow in a quantum state that is partly in the propagating regime (for example in the region B in Fig. 7.6) and partly in the evanescent regime (for example in the region C in Fig. 7.6). Consistency between them has to conserve the current at the point of propagating to evanescent crossover at the junction between regions B and C, there being no puzzle about the

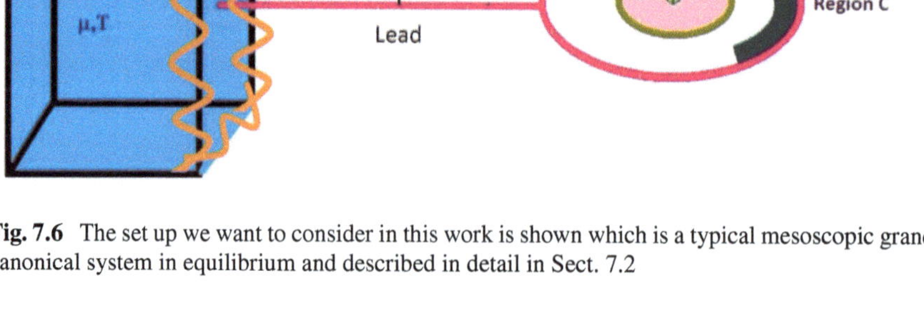

Fig. 7.6 The set up we want to consider in this work is shown which is a typical mesoscopic grand canonical system in equilibrium and described in detail in Sect. 7.2

current in region B. For resolving the puzzle of current in region C we have put forward an expression in Eq. 1.59 that follows from a different approach under the barrier using analytic continuity that does not lead to a straightforward probabilistic interpretation of quantum mechanical currents. Also such an analytic continuation does not preserve the structure of Hilbert space.

Apart from this consistency check, the primary advantage of theoretically studying such mesoscopic tunneling currents is that there are alternate theoretical formalisms in mesoscopic regime to verify the existence of currents inside the system under the barrier from propagating asymptotic states far from the barrier. That means there are theoretical cross checks for propagation under the barrier without studying the states under the barrier. Thus observations and calculation on current under the barrier can help us support or eliminate ideas. In short there are asymptotic states for the current carrying states in the ring and one can determine the current in the ring from these asymptotic states [8, 9], that is again similar to the Larmor clock. Now to calculate a scattering matrix element we may need an equation of motion but the derivation of Eq. 3.45 is insensitive to details like whether one is allowed to analytically continue the wave-function above the barrier to below the barrier or one should have a separate equation of motion for evanescent states. Rather as we have shown, all that Eq. 3.45 cares is analyticity of scattering matrix elements, it should be legitimate to analytically continue wave-functions inside the system as long as it does not destroy the analyticity of scattering matrix elements. Thereby we are claiming that Eq. 3.45 is more fundamental than Schrodinger equation as it gives a lot of new measurable quantities like the hierarchy of DOS that Schrodinger equation does not. So it can be used to make sense of tunneling electrons under the barrier, for which the Schrodinger equation cannot make any conclusive statement. In the limit of a closed system, Eq. 3.45 is also consistent with what one gets from Schrodinger equation. One can make the entire ring to be in the evanescent regime and the currents due to evanescent states have to be consistent with the asymptotic theory. A cartoon observer is shown in region A or lead of Fig. 7.6 and another cartoon observer is shown inside the ring. Observer A need not have any idea about the potential inside the ring. This observer does not know if the tunneling region extends over the entire ring or only a part of the ring. This observer only needs to know the infinitesimal change of $dU = \epsilon$ without any knowledge of U. He can measure the scattering phase shift (a theoretical measurement using analysis) in the wave-function in the lead and from there infer the current inside the ring from the analyticity of the scattering matrix elements. Observer A can vary the incident energy and check the analyticity of scattering matrix elements from Eqs. 5.2 and 5.3. The observer B in the ring can extend or shrink the tunneling region and also determine the tunneling current from the analytic continuation of the internal wave-function which is an unsettled issue in quantum mechanics. If the two measurements agree then the measurement by observer in the ring has to be theoretically correct.

In terms of practical measurements, there are problems associated with measuring tunneling currents in quantum mechanics and whatever we know about quantum measurements through an entanglement of sample states with the states of the detector does not apply for evanescent modes. One can also not measure tunneling currents classically as that will require the detector to be placed under the barrier and classical detectors cannot work under the barrier. These problems can be avoided in the above described mesoscopic set up. Because such currents can also cause magnetization that can be observed in a practical experiment for further validation. This magnetization can be measured without disturbing the state in the system and thus not invoking the unresolved issues of quantum measurement. In mesoscopic systems electrons are quantum mechanical but fields are classical. We can measure the magnetization due to the currents classically using hall magnetometers or SQUID remotely. This is why Fig. 7.6 shows a detector to measure magnetization, placed above the ring. Thus currents due to evanescent modes inside a ring can be measured non-invasively without encountering the above mentioned problems of quantum measurement. So it makes a lot of sense to study persistent currents due to evanescent modes inside the ring in the set up of Fig. 7.6.

In Fig. 7.7 we redraw the system in Fig. 7.6 in order to depict the parameters used and the wave-functions in the different regions for which the details can be seen in the figure caption. Here the reservoir is shown as an irregular thick line on the left. Region A of Fig. 7.6 is called region O and the three regions in the ring are labeled I, II, and III. The Schrodinger equation in 1D is

$$\frac{-\hbar^2}{2m_0}\frac{d^2\psi(x)}{dx^2} + U(x)\psi(x) = E\psi(x)$$

The solutions of Schrodinger equation in regions O (or lead), I, II, and III of Fig. 7.7 give the wave-functions in absence of magnetic field as

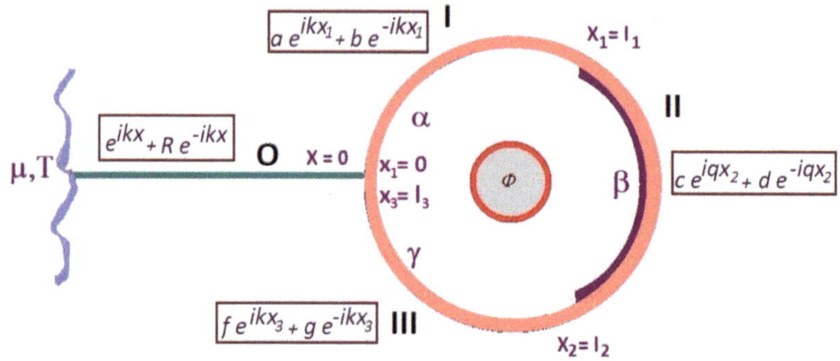

Fig. 7.7 The same ring as in Fig. 7.6 is depicted here to show the different parameters used in the theory. The ring region I is of length l_1, region II is of length l_2, and region III is of length l_3. The total length of the ring is $L = l_1 + l_2 + l_3$. The choice of coordinates with their origin is depicted in the figure. α is the AB phase for traversing region I, β is that for region II, and γ is that for region III, that appear in boundary conditions although the observable quantities like current can only depend on $\alpha + \beta + \gamma = \frac{2\pi\Phi}{\Phi_0} = \Phi'$ (say)

$$\psi_0 = \frac{1}{\sqrt{k}}(e^{ikx} + Re^{-ikx}) \tag{7.21}$$

$$\psi_I = \frac{1}{\sqrt{k}}(ae^{ikx_1} + be^{-ikx_1}) \tag{7.22}$$

$$\psi_{II} = \frac{1}{\sqrt{q}}(ce^{iqx_2} + de^{-iqx_2}) \tag{7.23}$$

$$\psi_{III} = \frac{1}{\sqrt{k}}(fe^{ikx_3} + ge^{-ikx_3}) \tag{7.24}$$

Here

$$q = \left(\frac{2m_0}{\hbar^2}(E - U)\right)^{1/2} \quad and \quad k = \left(\frac{2m_0}{\hbar^2}E\right)^{1/2} \tag{7.25}$$

We have discussed in Sect. 1.5 that the magnetic field can be included through the boundary conditions and the magnetic field results in a current in the ring. For $U = 0$ this becomes identical to the system of Fig. 1.3.

Now to consider the situation when $E < U$, we know from Eq. 1.59 that the probability current density in region II will be

$$J(k, \Phi) = \frac{e_0\hbar}{m_0 i}\left[dc^*e^{-i\Phi'} - cd^*e^{i\Phi'}\right] \tag{7.26}$$

Therefore differential current density in an interval dE will be

$$\begin{aligned} dJ(k, \Phi) &= \frac{e_0\hbar}{m_0 i}\left[dc^*e^{-i\Phi'} - cd^*e^{i\Phi'}\right]\frac{dn}{dE}dE \\ &= \frac{e_0}{hi}\left[dc^*e^{-i\Phi'} - cd^*e^{i\Phi'}\right]dE \end{aligned} \tag{7.27}$$

Here $\Phi' = \frac{2\pi\Phi}{\Phi_0}$. Now one can completely forget the factor $\frac{dn}{dE}$, if one uses the normalization constant given by the RHS of Eq. 1.10 and normalization constant is a freedom. We have so far used a normalization constant of $\frac{1}{\sqrt{k}}$ and an additional factor which is the square of $(\frac{2\pi m_0}{h^2})^{\frac{1}{2}}$ can be multiplied now (see Eq. 1.10). Instead if we solved the scattering problem with a normalization constant of unit incident flux then $\frac{dn}{dE} = \frac{1}{hv}$. Similarly, differential current density in region I where there is no potential will be

$$dJ(k, \Phi) = \frac{e_0}{h}\left[|a|^2 - |b|^2\right]dE \tag{7.28}$$

Thus if $E > U$ then the current in the regions I, II, and III will be given by $\frac{e_0}{h}(|a|^2 - |b|^2)$, $\frac{e_0}{h}(|c|^2 - |d|^2)$, and $\frac{e_0}{h}(|f|^2 - |g|^2)$, respectively, and this follows from the standard definition of quantum mechanical current even in the presence of flux Φ. Continuity of current implies that they should be all equal and that obviously comes out in our calculations. In absence of flux a, c, and f are the amplitude of clockwise moving electrons while b, d, and g are the amplitudes of anticlockwise electrons and they can only differ by a relative phase because none of them is preferred over the other. Flux breaks this symmetry so that $(|a|^2 - |b|^2)$ is the difference between the number of electrons moving clockwise and those moving anticlockwise in region I and that constitutes a current. This means that when Φ is zero then there is no current in the system as is expected for an equilibrium situation. However for $\Phi \neq 0$ the current is finite. It is an equilibrium current called persistent current, purely quantum mechanical in origin as discussed in Chap. 1. Equation 7.27 is far more complicated to be interpreted physically. Simple decoupling in terms of clockwise moving electron probability and anticlockwise moving electron probability is not possible. Time reversed amplitudes c^* and d^* appear in the expression. Tunneling under the barrier has long been studied as an avenue for superluminality but conclusive results do not exist as discussed at the beginning of this section. Current expression with current conservation conclusively shows that time reversed states have a role to play in the propagation.

This current is therefore flowing in the ring and magnetizing the ring. This magnetization can be measured by the detector on top and hence can be also experimentally verified. Without resorting to a practical experiment, this current expression for evanescent modes can be subjected to several tests. First is that it satisfies current continuity at the junction of regions I and II (see Fig. 7.7). Secondly an experimentalist can non-invasively measure and verify the numerical values of the current. As a third verification consider theoretically reducing the regions I and III to zero lengths, in which case the current and magnetization will be entirely due to evanescent modes. Regions I and III being reduced to zero any magnetization observed has to be purely due to current carried by evanescent modes. Such a magnetization of an entire ring carrying current due to tunneling will prove propagation under the barrier. Without waiting for an experimentalist to measure and verify this there can be an independent theoretical verification which works only in the mesoscopic regime as has been elaborated. An observer in the lead can remotely calculate this current without even knowing if the regions I and III are reduced to zero length or they have a finite length. For that we have a novel expression for current derived in [8, 9] and stated below.

$$dJ(E, \Phi) = \frac{c}{\pi} \frac{\partial \theta_R}{\partial \Phi} dE = \frac{2e_0}{h} \frac{\partial \theta_R}{\partial \Phi'} dE \tag{7.29}$$

Here $dJ(E, \Phi)$ is the differential current in an energy interval dE at a fixed flux Φ and $\Phi' = \frac{2\pi\Phi}{\Phi_0}$. The reflection phase shift is defined as $\theta_R = Arctan[\frac{Im(R)}{Real(R)}]$. The scattering phase shift $\theta_R(E, \Phi)$ is therefore from the asymptotic wave-function in the lead far away from the ring, and it is independent of the normalization constant chosen in the calculations. Observable current will be

$$I = \int_0^{k_f} dJ(k, \Phi) \tag{7.30}$$

Since our concern is what happens to tunneling electrons that seem to come out of the barrier before entering it [7], it makes sense to ask if there is density of states (DOS) under the barrier as well. Again standard approach is that this can be determined from internal wave-functions as well as asymptotic wave-function. However, the internal wave-function in tunneling regime is not consistent with the axioms of quantum mechanics because they cannot form a wave-packet, they are not elements of Hilbert space, and they are at most analytic continuation of wave-functions above the barrier. On the other hand, the asymptotic wave-function does not care if there is a state inside the ring or not and so if it gives the same value as that determined from analytic continuation of wave-function inside the ring, it can only mean two things. First, tunneling electrons do propagate under the barrier and secondly the asymptotic formalism settles a theoretical problem in the sense that analytical continuation implies physical continuation. So this is a theoretical experiment in the sense that what cannot be concluded from axioms of quantum mechanics can be concluded from the idea of a physical clock like Larmor clock. Of course currents are a better candidate as they can also be measured in a practical experiment and they do support this point of view as will be plotted and shown below. Nevertheless it does not harm to get further confirmation from DOS because after all DOS is a far more valuable quantity than just the currents. It is linked to all thermodynamic properties and not just currents. Besides, current conservation ensures certain features that can be settled purely at a level of counting with numbers without caring for a theory. DOS cannot be directly measured like one can measure current from magnetization, but is a much more complex concept connected to normalization, renormalization, regularization, etc., and if there is a scope to check purely theoretical consistency with the Larmor clock, that can be exploited by checking the following Eq. 7.31. For different systems in the propagating regime, we have already checked that in Chaps. 5 and 6 but now we have the DOS also leading to a current in the tunneling regime and current can be practically measured and so in our opinion worth studying in future very specifically. Usual wisdom is that currents are carried by quantum states and will qualitatively and quantitatively reflect the DOS. That is we expect

$$\frac{d\theta_R}{e_0 dU} = \pi\rho_d^u = \frac{m_0}{2\hbar^2 s}\Big[\frac{|c|^2}{2s}(1 - e^{-2sl_2}) + \frac{|d|^2}{2s}(e^{2sl_2} - 1) + (cd^* l_2 e^{i\Phi'} + dc^* l_2 e^{-i\Phi'})\Big] \tag{7.31}$$

Here ρ_d^u is the DOS in region II under the barrier and hence the superscript u, LHS giving the expression that can be obtained from asymptotic wave-function (see Eq. 3.57), while RHS is that obtained by integrating the LDOS from the analytic continuation of internal wave-function (see Eq. 3.56).

Let us first discuss the LHS of Eq. 7.31. Only for a constant potential of finite length, the integration in Eq. 3.52 can give a simple form as the RHS of Eq. 7.31. That the LHS of Eq. 7.31 will give $\pi\rho_d^u$ was generally established using BC. Now to check if there can be a state under the barrier, a constant potential is good enough.

It is to be noted that the LHS, that is $\frac{d\theta_R}{e_0 dU}$ as an expression for DOS does not explicitly use the tunneling wave-functions because its derivation is independent of whether the states are tunneling or propagating in region II. When we match boundary conditions as in Eqs. 1.53 and 1.54 to calculate scattering phase shifts we can just say that we are working with wavelets (Fourier components) of a wave-packet with a certain interpretation of probability and currents in quantum mechanics. For constant potentials we do need the Schrodinger equation in order to establish a relation between incident energy E, asymptotic wave vector k, and internal wave vector q. But now for the tunneling states it is unfair to say s is the internal wave vector. So we would rather say that the LHS is a general formula for ρ_d^u that applies for any potential, derived from BC using an infinitesimal change dU, without any knowledge of U. We can also say that θ_R has been calculated by assuming some kind of relation between k and s that does not destroy the analyticity of θ_R. Besides, phase change, or change in θ_R, is not determined by the exact relation between k and s, but is rather determined by the phase singularities and the topology of the complex plane.

The RHS of Eq. 7.31 uses the solutions of Schrodinger equation explicitly, but analytically continued. The RHS expression, consisting of 3 terms in first brackets, is completely different for evanescent modes than that for propagating modes and has to be found by integrating the absolute value of analytically continued wave-function in region II with appropriate normalization constant. So the RHS of Eq. 7.31 is positive definite. Interestingly, the RHS of Eq. 7.31 giving DOS under the barrier has some negative terms signifying there are partial states under the barrier for which counting in real numbers (measure) can yield a negative answer. Noting that ρ_d^u is positive definite, and also $s = (\frac{2m_0}{\hbar^2}(U - E))^{1/2}$ is positive, $[e^{2sl_2} - 1] \geq 0$ and $[1 - e^{-2sl_2}] \geq 0$, the remaining terms on RHS of Eq. 7.31 are also positive. Although the negative terms do not dominate over the positive terms, they do signify presence of processes back in time. However, there is no evidence that such processes can carry signal or information because for that we need propagating wave-packets. We have shown in Chap. 6 that propagating wave-packets can carry signal to the past using Fano resonances. Of course if we connect more leads to the ring being discussed now, then there may occur possibilities of transmission zeroes and Fano resonances and then these negative terms can manifest in some way. Below we will numerically check the equality in Eq. 7.31.

One may further say that there are states under the barrier given by Eq. 7.31 consisting of three terms in first brackets. But all three terms are not responsible for carrying current. Only the last term carries current as Eq. 7.27 confirms. An expression like e^{2sl_2} on RHS can grow indefinitely as l_2 is increased and to balance that the current carrying part of DOS that is $(cd^* l_2 e^{i\Phi'} + dc^* l_2 e^{-i\Phi'})$ becomes fewer and fewer. Current propagates only through these partial states. So if these partial states become fewer and fewer, then the electron takes lesser and lesser time to traverse the barrier. Only the electrons in these partial states show up in the Hartman effect but as these states do not constitute a propagating wave-packet they cannot transmit a signal.

We therefore proceed to demonstrate that Eq. 7.27 gives the same current as Eq. 7.29 and that the equality claimed in Eq. 7.31 is true. As we have discussed in Chap. 1 that one can set up the boundary conditions that lead to a set of simultaneous equations, that can be solved to obtain R, a, b, c, d, f, and g of Eqs. 7.21–7.24 in terms of known parameters like l_1, l_2, l_3, incident wave vector k, and applied magnetic field through the parameters α, β, and γ. In Fig. 7.8 we plot the current density versus kL which by current conservation has to be the same throughout the ring. For evanescent regimes we use Eq. 7.27 and for propagating Eq. 7.28, respectively. Three sets of parameter values have been used and mentioned inside the figure. Note that region II has a positive constant potential of $e_0UL = 49$. That means for k values up to $kL = 7.0$ the region II carries only evanescent states and so a vertical dotted line has been drawn at $kL = 7.0$ as a guide to the eye. The solid curve corresponds to a situation wherein $l_1 = l_3 = 0$ implying the entire ring has a positive potential $e_0UL = 49$, which means for kL values up to $kL = 7$, the entire ring carries evanescent modes only. The inset to Fig. 7.8 accentuates this range from $kL = 0$ to $kL = 7.0$. Above $kL = 7$ the states become propagating. The dashed curve and dash-dotted curve correspond to situations when the states in the ring are partially propagating and partially evanescent, wherein regions I and III are propagating while region II (see Fig. 7.7) will have evanescent modes up to $kL = 7$. Above $kL = 7$ the entire ring will carry propagating modes only. Current density is identically the same as calculated from the alternate formalism of Eq. 7.29 for all these situations, which means also for the situation of our interest that is the solid curve of Fig. 7.8 which confirms what we wanted to show. Our expression for current in Eq. 7.27 must be correct as it agrees with Eq. 7.29 which is a formalism that does not require us to treat the evanescent modes in any special way.

Another interesting feature to be noted from Fig. 7.8 is that as the region II of Fig. 7.7 or the region C of Fig. 7.6 is enlarged by making l_1 and l_3 shrink to zero keeping total length L unchanged, then for energy $E < e_0U$ current in the ring makes a gradual shift from paramagnetic to diamagnetic. This is shown in the inset to Fig. 7.8. When the length of region II is $l_2 = 0.5L$, then the current in the ring changes from diamagnetic to paramagnetic at $kL = 6.68$ (dash-dotted curve). When l_2 is increased to $0.7L$ then this crossover happens at $kL = 7.47$ (dashed curve). For the solid curve, this cross over occurs at $kL = 7.67$. This diamagnetic to paramagnetic oscillation in current is a property of propagating modes only, and when the entire ring is in the tunneling regime then under the barrier current can only be diamagnetic. This has to do with the fact that propagating modes have parity effect wherein alternate states alter in spatial symmetry. Even states are symmetric with an even number of nodes while odd states are anti-symmetric with an odd number of nodes. Evanescent mode wave-function is not wave like in nature at all. They have no nodes and so also no parity effect. This has been discussed in detail in Chap. 2. One may ask the question if there is an observable diamagnetic shift of the currents due to a partial or complete tunneling region inside the ring. Observable current given by Eq. 7.30 will be the area under the curves in Fig. 7.8 up to a Fermi wave vector k_fL. In Fig. 7.9 we have plotted this integrated current as a function of k_fL and at low energies we do see a diamagnetic shift up to $k_fL = 8.35$. But there is no such discernible diamagnetic

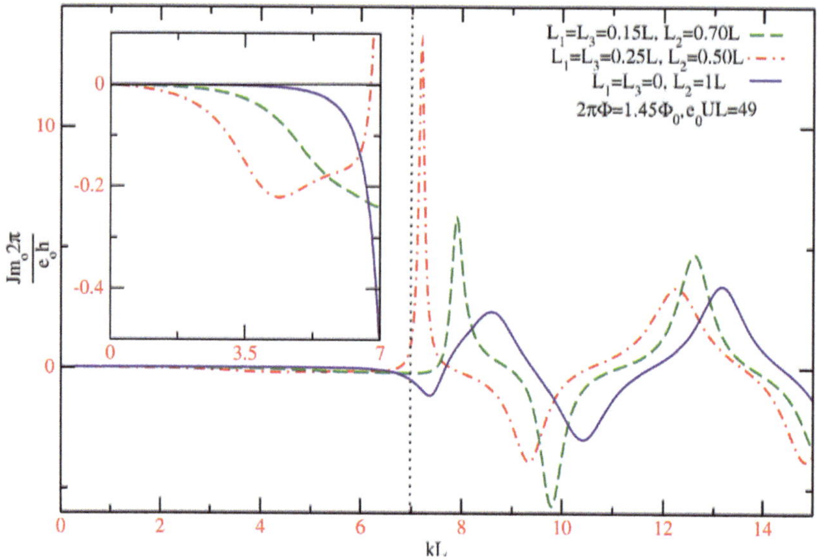

Fig. 7.8 Plot of probability current density in region II as a function of wave vector kL as obtained from Eq. 7.27 for three different systems corresponding to different lengths of regions I, II, and III for the system shown in Fig. 7.7. Current conservation implies that the current in regions I and II will be the same. Current in region II has to be also obtained from Eq. 7.29 and it confirms this. The vertical dotted line at $kL = 7$ gives the crossover point for the parameters used in this figure wherein modes in region II change from evanescent to propagating. We have taken $\hbar = 1$ and $2m_0 = 1$

shift at higher energies. But one can always have higher energy evanescent modes in the ring that show a diamagnetic shift at such energies. Such high energy evanescent modes are possible in presence of sub-bands due to lateral confinement as higher sub-bands have higher propagating thresholds. Interestingly this integrated current shows roughly same magnitude of current in the evanescent regime as that above the barrier. The peak value of the current at the first peak is roughly –1 while the propagating current peak values go up to ±2 (see Fig. 7.9). This is because in presence of partial or full tunneling the current is small in magnitude but always diamagnetic resulting in a larger area under the curves. Propagating state current alters between paramagnetic and diamagnetic and has a lot of cancelation effect. Corresponding graphs for Fig. 7.8 for multichannel ring have been published in [16]. Also the results of this section have been published in [17].

For completion we have also plotted DOS for the region II in Fig. 7.10. We plot the RHS and the LHS of Eq. 7.31 that coincide with each other for all three graphs as expected. To be noted that the LHS of Eq. 7.31 like the RHS of Eq. 7.29 is a general formula applicable to any arbitrary potential while the RHS of Eq. 7.31 is being calculated for a particular system to validate it. Curves are for the same value of parameters used in Fig. 7.8 and the same convention is followed. Note that all three curves show that when DOS under the barrier is high then current is also high.

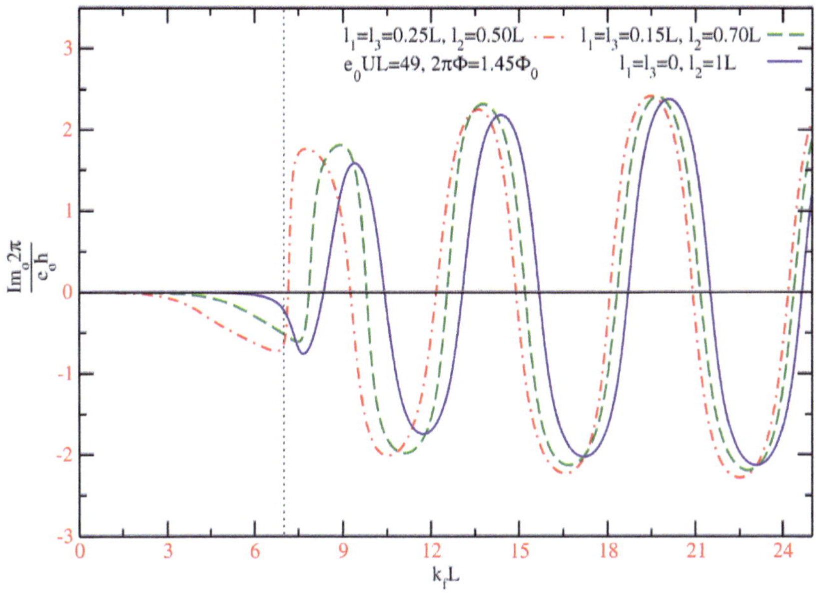

Fig. 7.9 Plot of measurable current in region II as a function of Fermi wave vector $k_f L$ as obtained from Eq. 7.30 for the same three different systems considered in Fig. 7.8. We have taken $\hbar = 1$ and $2m_0 = 1$

In summary, asymptotic theory works for current as well as DOS and has been also used to show that propagation under the barrier is a reality. Propagation under the barrier is not a well defined problem in quantum mechanics and so the asymptotic theory that goes beyond quantum mechanics uses the idea of physical clocks and analyticity of scattering matrix elements helps confirm this. This is referred to as a theoretical experiment but one can also make clean practical experiments to confirm this further. It also provides evidence that time reversed states have a role to play in propagation under the barrier. So there is a lot of merit in trying to address the question of propagation time and superluminality of tunneling particles. However, Eq. 6.15 cannot be used as a criterion because that requires construction of a wave-packet. Also note that the discussions related to point P in Fig. 5.7d are for propagating modes in 1D. The problem of tunneling time although found to be more complicated, the asymptotic theory can be used to understand currents and thermodynamic properties.

7.3 AC Response

Consider a sinusoidal voltage [3] of frequency ω, $U_\beta(\omega)$ applied at incident lead β. The current measured at lead α will be

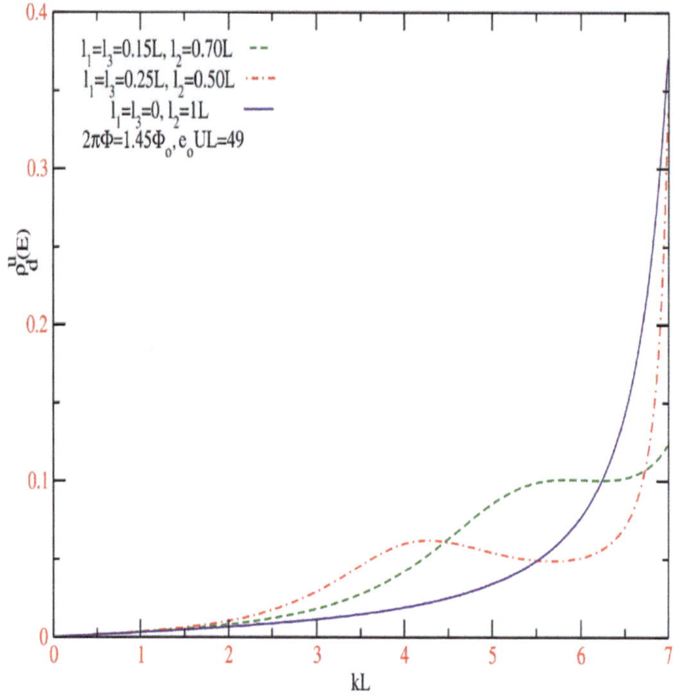

Fig. 7.10 Plot of DOS in region II only for the same set of parameter values used in Fig. 7.8 and Fig. 7.9. We have taken $\hbar = 1$ and $2m_0 = 1$

$$I(\alpha, \omega) = G_{\alpha\beta}(\omega)U_\beta(\omega) \tag{7.32}$$

where $G_{\alpha\beta}(\omega)$ is the dynamical conductance matrix and is given by

$$G_{\alpha\beta}(\omega) = G^0_{\alpha\beta} - i\omega E_{\alpha\beta} + K_{\alpha\beta}\omega^2 + O(\omega^3) \tag{7.33}$$

$G^0_{\alpha\beta}$ is the dc-conductance matrix. $E_{\alpha\beta}$ is proportionality constant for ω and governs the displacement currents and is given by

$$E_{\alpha\beta} = e_0^2\rho_{pd}(\alpha, \beta, E) - e_0^2 \int d\mathbf{r}'\rho_e(\alpha, \mathbf{r}', E) \int d\mathbf{r} g(\mathbf{r}, \mathbf{r}')\rho_i(\mathbf{r}, \beta, E) \tag{7.34}$$

where $g(\mathbf{r}, \mathbf{r}')$ is the effective interaction potential. The experimental situation explicitly involves β as the incoming channel and α as the outgoing channel.

7.4 Concluding Remarks

Several fundamental problems in physics have been addressed in this book and to several of them an answer has been given.

1. Time travel: This problem is encountered in different branches of physics. Newtonian mechanics suggests that the past is gone, future is yet to unfold, and present is an instant of time. Theory of relativity completely destroys this notion. In relativity, the past, present, and future are equally real and nothing is gone or yet to unfold. Relativity shows us that we live in a four-dimensional manifold (space-time) and the curvature of this four-dimensional manifold determines gravity. Since gravity behaves in a predictable way, say the orbit of the earth behaves in a predictable way, the curvature of space-time must be also fixed and predetermined. This means events in space as well as in time are all predetermined. We cannot use our free will to determine events in this four-dimensional space-time. Classically we cannot access the past but the past and future very much exist. Relativity also gives a very well established understanding of what actually is time. Or rather it gives a clear understanding of a time difference between two events in space-time. Essentially a body in free fall between two points in four-dimensional space-time will follow the path of least action, which means given the free fall of a body between two points in space-time we get a precise value for the time difference between these two events. Such a clear understanding of time was absent in quantum mechanics. Our book shows that we can have two Landauer type reservoirs that are completely classical. Time difference between two points in the two reservoirs is therefore precisely the classical time we get from relativity. Now one can send a quantum mechanical wave-packet from one point to another using mesoscopic phenomenon. A wave-packet is a pattern propagating in space so it can carry a signal or information from one classical point to another. If the wave-packet gets deformed due to dispersion then signal will also get deformed. But one can find parameter regimes where one encounters a Fano resonance and this resonant channel can transmit the signal undispersed. The signal can in fact be also transmitted in negative time which means to the past. Using propagating wave-packets to establish this result is completely consistent with the axioms of quantum mechanics.

2. Alternate view of quantum mechanics leading to a local theory: The standard view of quantum mechanics is that we start from 5 axioms. Then we establish the space in which quantum dynamics unfolds. That is the Hilbert space which is a special case of Banach space. One can establish isomorphism between the complex plane and the Hilbert space. A wave-packet made up of wavelets (Fourier components) is elements in this Hilbert space. These wavelets follow a differential equation called the Schrodinger equation that makes it extremely convenient to calculate numerical values of observables in different kinds of systems. Many different kinds of non-local behavior are thereby predicted by the Schrodinger equation and it may be fair to say that most are in agreement with observations. But are there local parameters that lead to such non-local behavior is a problem in physics that several great minds have worked on. One can mention a few of them to be Bohmian mechanics, diffusion model of quantum mechanics, etc. A simple example of non-local behavior is if we take two barriers of say transmission probability of T_1 and T_2 and place them in series then the resultant transmission probability is not $T_1 + T_2$. Our alternate view can be explained as follows. We do not need to establish a Hilbert space at all. We can say that the wave vectors of the wavelets making a wave-packet can be indexed by real numbers (it is to be noted that box normalization requires them to be indexed by natural numbers which will immediately lead to Banach-Tarski paradox and is completely wrong). There is a linear bijective map (say the scattering matrix but it can be simply seen as a map) that maps the incoming wavelets to the outgoing wavelets. This map is a continuous map or rather an analytic complex function. Functional analysis tells us that we should not try to count the wavelets in real numbers but rather try to measure them using the complex function that maps the incoming wavelets to the outgoing ones. Thus functional derivatives of this map or function help us derive entities like LPDOS, PDOS, injectivity, emissivity, etc., that cannot be calculated from Schrodinger equation. Schrodinger equation always requires a particular potential to be specified. The functional derivatives do not require the potential to be specified but just require that whatever be the potential, it can be deformed infinitesimally and continuously. LPDOS is for example a completely local quantity because by adding them up for all points we get PDOS. Adding PDOS for relevant channels we get injectivity and emissivity. Similarly, time taken by a classical particle from a point in one classical reservoir to a point in another classical reservoir through a quantum process can be seen as a sum of time spent (calculated quantum mechanically) at all intermediate points. We can use Schrodinger equation for a particular potential to calculate injectance and this injectance can also be obtained from the LPDOS.

3. State of a tunneling particle under the barrier: This problem is very ill defined in quantum mechanics and so our alternate view of quantum mechanics helps us conduct a theoretical experiment to observe some important aspects of this problem. These observations also suggest that our approach works even in regimes where quantum mechanics does not work. Our observations are consistent with analytical continuation of quantum mechanics.

4. Applications:

- First of all as stated in point 2 that to determine LPDOS, PDOS, etc., we do not need to know the exact potential inside a quantum system but we just need to know the effect of small changes in the potential. This can help experimentalists to determine these quantities without knowing the internal details of the system. Specially because we have seen that both scattering cross section and scattering phase shifts can be experimentally observed. This can help us characterize mesoscopic samples.
- If one goes to a doctor because he is suffering, the doctor does not think in terms of the sufferings. He will think in terms of heart, lungs, kidney, etc., and try to diagnose a local cause that has a global manifestation as suffering. Quantum mechanics is non-local which makes it difficult for engineers to think and analyze in terms of local entities. Wave-function depends on coordinates and quantum

numbers. Quantum numbers are global parameters. Absolute value of the wave-function can give local probability only if normalized by a global probability. Absolute value completely throws away the information contained in the phase of the wave-function. Besides, as the local probability thus defined depends explicitly on the local coordinate as well as the quantum numbers, its applicability is greatly reduced by the uncertainty principle. In case the wave-function is not normalizable, like in case of scattering states, we resort to regularization, but could not define local entities in terms of which we can analyze global entities. We tried other approaches like pair correlation function or density-density correlation function, etc., and none gives a global description of a quantum system in terms of local entities. Pair correlation function means we take the many body wave-function and integrate out (or average over) all coordinates except two and then fix one and see variation with the other. Density-density correlation function or current-current correlation function makes sense only if ensemble averaging is done. This ensemble averaging is not really phase space averaging but something that mimics a phase space averaging. For example one way is to add artificial terms in the Hamiltonian and averaging is done by varying these artificial terms and if possible then finally make these artificial terms tend to zero. We can see in Eqs. 7.1–7.4 as well as in Eqs. 7.5, 7.17, 7.29, and 7.31 that global quantities like currents in the sample or conductance of the sample can be written down in terms of local entities referred to as DOS hierarchy members. They do not explicitly depend on quantum numbers. They explicitly depend on coordinates that are the physical coordinates defining the sample. Some of them like injectivity or emissivity are partially averaged while PLDOS is a purely local entity. The same quantities appear in different combinations in the different formulas. They are like the heart, lungs, and kidney of a sample in terms of which an engineer can think to design a device. They determine mesoscopic response. The meaning of these local entities was not clear as they can be derived by appealing to semi-classical ideas like spin precession and Larmor frequency. We have made them general physically as well as mathematically by invoking Burgers circuit and analyticity.

- Formalizing quantum mechanical currents under the barrier can help develop new technologies that use quantum principles. The main challenge in this regard is that quantum phenomenon is difficult to control and harness because of quantum fluctuations and decoherence. Evanescent states are not as sensitive to fluctuations and decoherence like the propagating modes. Since these are exponentially decaying modes, their fluctuations are also damped but the signal is not much damped.
- Negative local partial density of states can lead to attraction between two electrons.

5. Can our point of view lead to unification of classical and quantum laws? The counter argument to no free will of point 1 above is that relativity does not account for creation and destruction of matter, which means we can always burn fuel to change curvature of space-time. But can we use our free will to burn fuel as we like (we caution that people should not extend our theory to interpret global warming. Besides it was proved a long time ago in neuro science that there is no free will. One can search Wikipedia to see the works of Benjamin Libet and related later works. So we are not responsible for irresponsible people)? Well six billion people on earth, exercising their free will over centuries, should have surely made it impossible for us to figure out how was the orbit of earth a few centuries ago. We in fact know how it was for thousands of years. Any burning of fuel essentially means converting some potential energy into kinetic energy. Our theory says that the difference between two classical times is related to changes in potentials in quantum mechanics. So we cannot use our free will to change potential at any point because that has to respect the time difference between two points or events in our four-dimensional space-time. May be low energy mesoscopic phenomena is the right thing to explore in order to unify classical and quantum laws.

References

1. Satpathi U, Singha Deo P (2016) Ann Phys 375:491
2. Datta S (1995) Electronic transport in mesoscopic systems. Cambridge University Press. Reprinted in 1999
3. Buttiker M (2001) Time in quantum mechanics. Springer, pp 279–303
4. Meena K, Singha Deo P (2022) Advances in theoretical and computational physics 5:458
5. Bandopadhyay S, Jayannavar AM (2004) arXiv:quant-ph/0411161 and references therein
6. Landauer R, Martin Th (1994) Rev Mod Phys 66:217–228 and references therein
7. Hartman TE (1962) J Appl Phys 33:3427
8. Akkermans E, Auerbach A, Avron JE, Shapiro B (1991) Phys Re Lett 66:76
9. Mello PA (1993) Phys Rev B 47:16385
10. Mailly D, Chapelier C, Benoit A (1993) Phys Rev Lett 70:2020
11. Reulet B, Ramin M, Bouchiat H, Mailly D (1995) Phys Rev Lett 75:124
12. Deblock R, Bel R, Reulet B, Bouchiat H, Mailly D (2002) Phys Rev Lett 89:206803
13. Deblock R, Noat Y, Reulet B, Bouchiat H, Mailly D (2002) Phys Rev B 65:075301
14. Levy LP, Dolan G, Dunsmuir J, Bouchiat H (1990) Phys Rev Lett 64:2074
15. Webb RA, Washburn S, Umbach CP, Laibowitz RB (1985) Phys Rev Lett 54:2696
16. Mukherjee S, Singha Deo P (2013) Physica E 47:86
17. Meena K, Singha Deo P (2023) Physica E 49:115680

MIX
Papier aus verantwortungsvollen Quellen
Paper from responsible sources
FSC® C105338

If you have any concerns about our products,
you can contact us on
ProductSafety@springernature.com

In case Publisher is established outside the EU,
the EU authorized representative is:
Springer Nature Customer Service Center GmbH
Europaplatz 3, 69115 Heidelberg, Germany

Printed by Libri Plureos GmbH
in Hamburg, Germany